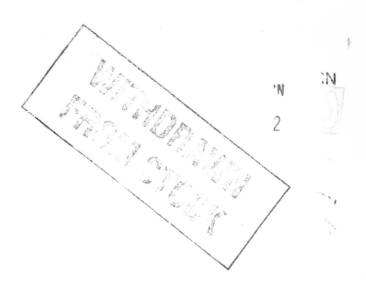

WITHDRAWN
FROM STOCK

'N

2

SEP 1500

URN

Fluorescence Spectroscopy

AN INTRODUCTION FOR BIOLOGY AND MEDICINE

Fluorescence Spectroscopy

AN INTRODUCTION FOR BIOLOGY AND MEDICINE

Edited by

AMADEO J. PESCE
Department of Medicine
Michael Reese Hospital and Medical Center
and
Department of Biology
Illinois Institute of Technology
Chicago, Illinois

CARL-GUSTAF ROSÉN
Department of Biochemistry
University of Stockholm
Stockholm, Sweden

TERRY L. PASBY
Department of Biochemistry
University of Illinois
Urbana, Illinois

Marcel Dekker Inc.   New York   1971

MARCEL DEKKER, INC.
95 Madison Avenue, New York, New York 10016

LIBRARY OF CONGRESS CATALOG CARD NUMBER: 76-154611
ISBN NO.: 0-8247-1539-X

PRINTED IN THE UNITED STATES OF AMERICA

299274

Dedicated to Dr. Gregorio Weber

Preface

Fluorescence spectroscopy as a tool has been met with a very rapidly growing interest among biologically oriented scientists in recent years. The sensitivity and versatility of this technique makes it useful in an increasing number of areas, where the main object is to solve primarily biological problems. Every year a large and increasing number of papers describing such applications appear in biochemical and medical journals, and a number of reviews have been written. It is possible to use fluorometry as a routine tool without knowing much about the theoretical background of fluorescence; but with the application of fluorescence spectroscopy in ever expanding new areas, there follows a growing need for understanding of the physicochemical laws governing the studied processes. In the application-type papers reference is frequently made to texts which, although they may discuss the phenomena in an excellent way, are more or less impossible to read for a person lacking a regular training in physics as so many biologists do (including some of the authors of this book). Besides, the fundamentals of fluorescence spectroscopy as used in biological work are scattered among so many different papers and journals, where different nomenclature has been used, as to make it a very tedious job to extract the necessary background information from the original literature.

The present text attempts to give a unified picture of the theoretical concepts and mathematical expressions, which are essential to anyone who wants to make full use of the powerful tool of fluorescence spectroscopy in solving problems of biological interest. In trying to avoid the mistake, made in some earlier books of a similar kind, of leaving information gaps between the separate chapters, we have rather permitted some overlap. An advantage of this is that each chapter may in general be read as a separate unit while it is still an integrated part of the total text. The authors believe that a biologist lacking advanced training in physics or mathematics will be able to extract from Chapters 2 through 7 most of the information on fluorescence spectroscopy which he will need in his work. Since no discussion of fluorescence phenomena is possible without reference to the underlying electronic events, Chapter 1 has been included as a general introduction to basic quantum mechanics and the theories behind chemical bonds and electronic transitions. A quantitative treatment of fluorescence spectroscopy requires the use of mathematics, but in this text we have attempted to reduce the mathematical expressions and derivations to simple terms. The reason for including the extensive derivations of polarization equations is that these derivations have not earlier been brought together in one place, and some of them have not even been published before, at least not in a form digestible to a nonmathematician. However, the chapter on polarization of fluorescence may very well be read without those derivations, and consequently they have been collected in an appendix.

Since the object of this book is to present a unified picture of the fundamentals of fluorescence spectroscopy rather than being a reference source for applications, examples of applications have been included in the text only to the extent that they serve to illustrate the phenomena and principles. However, the interested reader will find an abundance of references to such applications at the end of each chapter.

This book is the result of a joint effort of several people, who's study of fluorescence phenomena has been greatly stimulated and partly guided by Professor Gregorio Weber at the University of Illinois. The chapters were originally written by separate authors:

> Chapter 1 — Anton F. Schreiner
> Chapter 2 — Amadeo J. Pesce
> Chapter 3 — Terry L. Pasby
> Chapter 4 — Earl N. Hudson
> Chapter 5 — Carl-Gustaf Rosén
> Chapter 6 — Terry L. Pasby
> Chapter 7 — Diane Davenport

They were then arranged and revised so as to conform with each other and with the intentions of the main authors, who hope that in this shape the book will prove to be of value in helping biologists make fluorescence spectroscopy a useful tool in their work.

When the manuscript of this book was in its final stages, the team of authors responsible for this book lost a most inspiring member, and the profession of biochemical fluorescence spectroscopy lost one of its most promising young scientists. Dr. Terry L. Pasby died after a short time of acute illness on August 29, 1969. Although he was only 26 at the time of his death, his knowledge and ability in physical biochemistry were extraordinarily impressive, and he was already a full master of the topic presented in this book, which would not have been written but for his whole-hearted commitment to the work. His own research, in which advanced fluorescence spectroscopy was applied to protein binding and conformation studies, will be published with Dr. Gregorio Weber.

Chicago, Illinois   Amadeo J. Pesce
Stockholm, Sweden   Carl-Gustaf Rosén
February, 1971

Contents

List of Symbols

A	Molecule in the ground state; acceptor molecule
A^*	Molecule in the excited state
ΔA	Difference in instrument response for absorption
$\%A$	Percent absorption of light ($1 - \%T$)
c	Concentration of absorbing substance
c'	Constant of proportionality
D	Donor molecule
l	Standard error
E	Energy
E_s	Energy of molecule in excited singlet state
E	Einstein (mole of photons)
E	Electric vector of light
e	Electric charge of electron
$f_{\bar{\nu}}$	Fraction of light transmitted by monochromator at $\bar{\nu}$
$\bar{f}$	Oscillator strength
f	Fraction of bound ligand
F	Fluorescence intensity
F_b	Fluorescence yield of fully bound ligand of proteins
F_f	Fluorescence yield of free ligand in proteins
F_{rel}	$F_{\text{sample}}/F_{\text{standard}}$
F_{sample}	Fluorescence intensity of a protein ligand solution
F_m	Fraction of light absorbed by monochromator
ΔF	Difference in detector response for emission wave number (λ^{-1}) (reciprocal cm)
g_e	Degeneracy of lower state
g_u	Degeneracy of upper state
G	Ground state
$\mathcal{H}$	Orientation factor
H	Magnetic vector of light
h	Planck's constant
$h\nu$	Energy of light
I	Intensity of light
$I_\perp$	Perpendicular polarized light
$I_\parallel$	Parallel polarized light
$I(\lambda)$	Intensity of light at wavelength λ
$I(\lambda)_f$	Intensity of emitted light at wavelength λ

j	Apparent order of binding reaction
K_a	Association constant
K_d	Dissociation constant ($1/K_a$)
k_f	Rate constant for fluorescence emission
k_i	Rate constant for radiationless energy loss
k_x	Rate constant for intersystem crossing
KE	Kinetic energy
l	Optical path through sample
L	Ligand
(L)	Concentration of unbound ligand
(L_0)	Total concentration of ligand
M	Constant of proportionality, absorbance constant extinction coefficient
m	Mass
N	Avagadro's number; maximum number of ligand molecules bound by protein molecule; number of binding sites on protein
n	Refractive index
n	Quantum number
$\bar{n}$	Average number of binding sites filled by ligand
nm	Nanometer (10^{-9} meter)
P	Energy
p	Momentum (mv); **polarization**
P_b	Polarization of bound dye
P_f	Polarization of free dye
P_{obs}	Observed polarization
$P_{\bar{\nu}}$	Output/quanta input of photomultiplier at wave number $\bar{\nu}$
PE	Potential energy
P	Protein
(P)	Concentration of unbound protein
(P_0)	Total concentration of protein in solution
(PL)	Concentration of protein-ligand complex
q	Quantum yield
q_0	Unquenched quantum yield
Q	Quencher
Q^*	Excited quencher
$\bar{q}$	Charge
q	Quanta of light
r	Internuclear distance
R_0	Distance at which probability of energy transfer is 50%
R_{AB}	Distance from a to b
$R_{\bar{\nu}}$	Response of photomultiplier at wave number $\bar{\nu}$
S	Spin of electron
S_1	Singlet state
$S_{\bar{\nu}}$	Sensitivity of spectrometer at wave number $\bar{\nu}$
T	Triplet state
$\%T$	Percent transmission of light
V	Molecular volume
v	Vibrational frequency
$v = 0$, etc.	Ground state vibrational levels
$v' = 0$, etc.	First excited state vibrational levels
W_A	Energy of oscillator A

$W_{\bar{\nu}}$	Band width in wave number $\bar{\nu}$
α	Absorption coefficient of walls
∇^2	Second derivative $(d^2/d_x{}^2)$
∂	Torque

Chapter 1

Background and Physical Principles

Fluorescence spectroscopy rests upon several basic principles. One needs to be familiar with them in order to interpret both electronic absorption and emission spectra. This familiarity derives from examining the nature of light, molecular ground and excited state properties, and electronic absorption and emission mechanisms. The reader who wants a purely descriptive explanation of fluorescence spectroscopy is advised to start with Chapter 2. For those readers who want to fully understand the quantitative aspects of fluorescence and fluorescence polarization, we have included the following brief presentation of the physical principles. These subjects will be developed by semi-quantitative descriptions of underlying principles and by elucidating sample spectra encountered in practice.

The relationships between light and matter can be understood by employing wave theory and wave mechanics, one form of quantum chemistry. For example, we consider light as a wave. Furthermore, the properties of electrons of a molecule can be described in space and time by wave functions. The manipulations of mathematical forms of waves and wave functions will then describe what happens when light interacts with a molecule. In addition, we shall introduce the concept of molecular orbitals to describe the electrons of a complicated molecule. Using this molecular orbital model we shall then observe which electronic orbitals are changed when the molecule absorbs light, what characteristics of the molecule are changed, and how these changes relate to common absorption and emission spectra.

1.1 Light

A. Structure of Light

Light is a form of energy which has both wave and particulate properties. Therefore, either of two physical models may be used to describe experimental observations. The choice depends on the type of experiment, and the wave model applies to electronic spectroscopy. The latter model may be described in the following manner. One property of a wave is its wavelength λ, which is the distance from the crest of one wave to the crest of a neighboring wave in a continuous wavetrain (see Fig. 1.1). The wavelengths of light in the visible and ultraviolet (uv) regions of spectra, the regions of primary importance in (electronic) fluorescence spectroscopy, are in the approximate range 200 to 800 nm (nm $= 1 \times 10^{-7}$ cm $= 1 \times 10^{-9}$ m).*

Other characteristics of light are its velocity $c = 3 \times 10^{10}$ cm/sec and its frequency ν (sec^{-1}). Equation (1.1) relates wavelength, frequency, and the velocity of light:

$$c = \lambda \nu \tag{1.1}$$

ν can be thought of as the number of waves passing a point in one second. The use of Eq. (1.1) is demonstrated in the following example.

Visible light of wavelength 500 nm has frequency

$$\nu = \frac{c}{\lambda} \frac{3 \times 10^{10} \text{ cm/sec}}{500 \text{ nm}}$$

But,

$$1 \text{ nm} = 1 \times 10^{-9} \text{ m} \quad \text{or} \quad 1 \times 10^{-7} \text{ cm}$$

$$\nu = \frac{3 \times 10^{10} \text{ cm/sec}}{500 \times 10^{-7} \text{ cm}} = 6 \times 10^{14} \text{ sec}^{-1}$$

Let us now consider the particulate nature of light. The amount of energy per wavepacket (photon) of any frequency is given by the relationship

$$E = h\nu \tag{1.2}$$

where E is the energy in ergs; h is Planck's constant, 6.6×10^{-27} erg $\times$ sec. This is the relationship (1.2) pertinent when considering light as

---

*The most recently suggested terminology for 1×10^{-9} meter is the nanometer (1 nm $= 1 \times 10^{-9}$ meter $= 1$ mμ); it is in very little use compared to the term millimicron, but nm will be employed for this wavelength unit throughout the book.

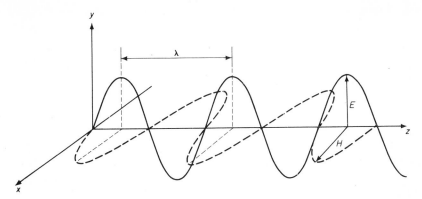

Fig. 1.1. One electromagnetic light wave with electric field vector **E** in the yz plane (yz polarization) and magnetic field vector **H** in the xz plane.

particulate energy, i.e., light is composed of "quanta." We use (1.2) by continuing with the example above, i.e., the energy associated with the wavetrain of 500 nm is

$$E = (6.6 \times 10^{-27} \text{ erg} \times \text{sec}) \times (6 \times 10^{14} \text{ sec}^{-1})$$

$$= 3.96 \times 10^{-12} \text{ ergs}$$

This packet of light, a photon, is the quantum of energy at this wavelength (500 nm). It is indivisible, and when absorbed by a molecule the entire photon with all its energy is absorbed. In other words, when electronic transitions occur, *the photon is not dissected by its interaction with the molecule such that one part is absorbed and another part continues as a photon with reduced energy.*

It is more convenient to employ larger energy units, and one way to accomplish this is to speak of interactions on a molar basis, viz., a mole of photons, 6.023×10^{23}, is defined as an Einstein (E). For example, the energy of one Einstein at wavelength 500 nm is equal to

$$E = (6.02 \times 10^{23} \text{ units/mole}) \times (3.96 \times 10^{-12} \text{ erg/unit})$$

or

$$E = 24 \times 10^{11} \text{ ergs/mole}$$

Since

$$1 \text{ kcal} = 4.18 \times 10^{10} \text{ ergs}$$

$$E \sim 60 \text{ kcal/mole}$$

This would be the amount of energy absorbed, if each molecule of one mole of molecules would absorb one photon of light having wavelength 500 mμ.

The interaction of the oscillating electric field component of light with (oscillating) charges in the molecule (e.g., electrons) is the basis of the spectroscopic phenomena which we shall discuss.

B. SEVERAL INTERACTIONS OF LIGHT WITH MOLECULES

Polychromatic light is composed of a mixture of electromagnetic waves of many wavelengths. Each wave can be thought of as oscillating electric and magnetic field vectors. Figure 1.1 shows the wave equivalence of one photon having wavelength λ and propagating in the z direction. Electric and magnetic field vectors are designated **E** and **H**, respectively (boldface type indicates vector quantities, viz., they have magnitude and specified direction). Several waves with the same or different wavelengths may be in the yz plane (light is yz-polarized) or in several planes (nonpolarized). We shall now describe two interactions of light with matter.

Consider the results of **E** interacting with an electron dipole (e.g., a molecule with relative positive and negative ends). In other words, the uniform field **E** (the wavelength is much larger than the molecule) of certain electromagnetic radiation can interact with a dipole having two charges $+q_A$ and $-q_B$ separated by distance $\mathbf{R}_{AB}$. One result may be that the dipole will tumble, since it experiences the torque δ caused by this interaction of external field and dipole moment $\boldsymbol{\mu}$ (see Fig. 1.2). The torque ($\boldsymbol{\delta}$) is defined as

$$\boldsymbol{\delta} = \boldsymbol{\mu} \times \mathbf{E} \qquad (1.3)$$

and the dipole moment is defined by $\boldsymbol{\mu} = Q\mathbf{R}_{AB}$ (charge times distance). This, for example, is a simple way to visualize the interaction of

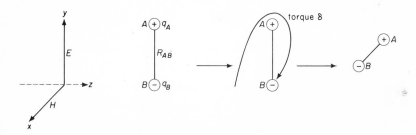

Fig. 1.2. Electric field vector **E** interacting with dipolar molecule A–B causing torque $\delta = q\mathbf{R}_{AB} \cdot \mathbf{E}$ (charge $\times$ distance $\times$ electric field).

microwave radiation with a molecule having a permanent dipole moment.

Consider a second type of interaction, i.e., whereby light excites electrons in molecules. Both electric (**E**) and magnetic (**H**) components of the light can actually interact with mobile electrons. Let us again assume that the light is yz-polarized and defined by the two vectors $\mathbf{E}_{yz}$ and $\mathbf{H}_{xz}$. Because of its (static or moving) charge and orbital motion, the electron can interact with both **E** and **H** of the radiation, respectively. For example, the orbital motion of the electron produces a small magnetic field $\mathbf{H}_e$ and magnetic moment $\boldsymbol{\mu}_m$, just as an electric current in a coil of wire is able to produce a similar field (Fig. 1.3). The orbital moment has classical magnitude

$$\boldsymbol{\mu}_m = \frac{er\mathbf{v}}{2c} \tag{1.4}$$

where e, r, and $\mathbf{v}$ are electron charge, radius of motion, and velocity, The text of Eyring et al. (1944) can be consulted for the quantum mechanical equivalence of (1.4). There is an additional small magnetic field ($\mathbf{H}_s$) generated by the spin of the electron about its own axis (Fig. 1.3). This effect is of no consequence here.

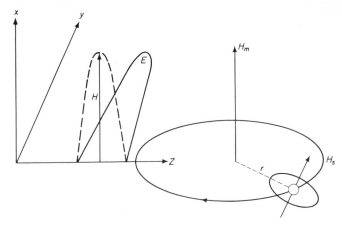

Fig. 1.3. Magnetic fields of the electron due to orbital and spin-axis motions.

The relative electric and magnetic contributions to the total force **F** acting on the electron moving in an electromagnetic field can be determined by examining Eq. (1.5)

$$\mathbf{F} = [e\mathbf{E}] + \left[\frac{e}{c}\mathbf{v} \times \mathbf{H}\right] \tag{1.5}$$

where e and $\mathbf{v}$ are the charge and velocity of the electron, and c is the light velocity $(3 \times 10^{10}\,\mathrm{cm/sec})$. The first and second terms represent contributions from electric and magnetic fields of light, respectively. The first term is the force exerted by an electric field on an electron and the second the force on the magnetic moment of the electron by the magnetic field. However, the second term of (1.5) is very small, i.e., velocities of electrons in light atoms are much slower than the velocity of light. Therefore, when a molecule interacts with light, the force on the electron arises almost entirely from the electric field interaction. Only this electron–electric field interaction will be discussed here since it is the most important phenomenon operative in electronic absorption and emission spectra of molecules in this text.

There is another frequently used physical picture of light absorption for electronic transitions in molecules. This derives from visualizing molecules as oscillating (vibrating) dipoles which absorb energy from the oscillating electromagnetic radiation during the "resonance condition." Absorption by molecules without permanent dipole moments can also be rationalized by the model, since it is known that an electric field (e.g., from the light) can induce a dipole moment. These descriptions are pictorial and simple but serve to visualize ways in which molecules can interact with the electric field component of light.

Although molecules can get into excited (electronic) states by interacting with electromagnetic radiation, light of only very specific wavelengths will be absorbed or emitted by a particular molecule. In other words, *molecules have a very high probability of absorbing incident light of only very specific wavelengths* from which absorption bands arise (see Sec. 1.5). This specificity exists because molecules can only be in discrete energy states. Transitions can take place from a state of low energy to one of higher energy by absorption of radiation.

Similarly, the loss of absorbed energy by the emission of light is also more probable at certain wavelengths. The quantum mechanical treatment given below accounts for this quantization and the perturbation (interaction) of the molecular electronic energy by light.

1.2 Quantum Chemistry

A. Some Principles of Quantum Chemistry

If one knew all forces acting on every electron and nucleus during the irradiation time of a molecule by light, exact quantitative predictions of molecular properties would be feasible. *At best*, one can make only time-average predictions. Even the latter predictions are only approximations, because explicit mathematical treatments of molecular electronic properties are also approximations. Mathematical difficulties exist because all electrons and nuclei interact among themselves and one with the other at the same time. For example, to calculate the forces on one electron of a carbon atom one would have to consider the force of the nucleus on this electron and the force of the other five electrons upon the one described. In order to understand how the other five effect that electron, the interactions of each of these electrons with the nucleus and with each other must be considered. As expected, the complexity increases with increasing numbers of electrons and nuclei in molecules, and to simplify matters certain assumptions or models are used which approximate these complex interactions.

As stated above, certain assumptions and models are used to approximate the time average values of molecular properties such as dipole moments, i.e., a molecule can be represented in a wave function according to the theory of quantum mechanics. Another principle of quantum mechanics is that all molecular properties, such as the energies of absorption and emission bands, intensities of these bands, and the energy of a molecule or its dipole moment, can be predicted by using the mathematical form of this wave function. These wave functions are normally denoted by the symbol ψ. Another quantity, the *operator equivalent* of a molecular property, is needed to complete the description. In the next section we present the simple case of the free electron, the use of wave functions, and the operator equivalent of energy, H.

B. SIMPLE DEVELOPMENT OF THE SCHRÖDINGER OR WAVE EQUATION

It was suggested by de Broglie (1925) that both light and matter have simultaneous wave and particulate properties. Using the equations

$$E = mc^2 \tag{1.6}$$

$$E = h\nu \tag{1.7}$$

$$\lambda = \frac{c}{\nu} \tag{1.8}$$

it is evident by substitution that

$$\lambda = \frac{h}{mc} \tag{1.9}$$

m and E are mass and energy of the same quantity of matter, h is the Planck constant, ν is the frequency (sec^{-1}), and c and λ were defined above. For velocities slower than c (3×10^{10} cm/sec) de Broglie suggested that the velocity v replace c in Eq. (1.9), which becomes

$$\lambda = \frac{h}{mv} = \frac{h}{p} \qquad (1.10)$$

where $p = mv$ is the momentum.

Since the momentum p and wavelength λ are associated with mass and waves, respectively, Eq. (1.10) expresses the dual nature of either matter or light. This has been verified experimentally (Davisson and Germer, 1927), and descriptions of this experimental evidence may be found in any good physical chemistry textbook (Glasstone, 1946).

Let us consider the electron as the particle in question. Without directing forces acting on it, its energy is entirely kinetic, or

$$KE = \tfrac{1}{2} mv^2$$
$$= \frac{1}{2} m \left(\frac{h}{m\lambda}\right)^2 \qquad (1.11)$$
$$= \frac{h^2}{2m\lambda^2}$$

Therefore, *the kinetic energy of the free electron is related to an associated wavelength* (h and m are constants).

The behavior of the electron as a wave led to the intuitive conclusion that it be described mathematically by known expressions, which were previously used to describe the behavior of physically observable waves. We may let these observable waves, for example, be the ones produced by plucking a cord which has its two ends tied to two walls opposite each other (Fig. 1.4). In fact, for certain states called stationary states, a particle (electron in our case) can be represented by a

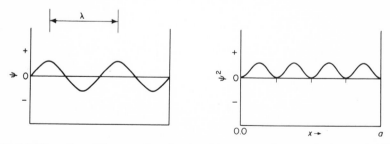

Fig. 1.4. The value of the wave function ψ and its square ψ^2 versus displacement x for the function $\psi = a \sin(2\pi(x/a))$.

standing wave. The equation for such a wave is

$$\phi = \left(a \cos 2\pi \frac{x}{\lambda}\right)\left(\cos 2\pi \frac{t}{p}\right) \tag{1.12}$$

This equation is the product of two terms; one is a function of time t and the other of displacement x. Abbreviated, Eq. (1.12) may be written as a product

$$\phi = (\psi_x)(\psi_t) \tag{1.13}$$

where

$$\psi_x = a \cos 2\pi \frac{x}{\lambda} \tag{1.14}$$

$$\psi_t = \cos 2\pi \frac{t}{p} \tag{1.15}$$

and a is a constant, p the period, ϕ the displacement vector, x the direction of wave progress, and t the time. Therefore, *the wave equation* ϕ is the product of a space-coordinate function ψ_x and a time-dependent function ψ_t. Using $\psi = \psi_x$ after evaluating $d^2\psi/dx^2$, solving for $1/\lambda^2$, and substituting the latter expression into (1.11), one obtains

$$KE = -\frac{h^2}{8\pi^2 m}\frac{d^2\psi}{dx^2}\frac{1}{\psi} \tag{1.16}$$

Total energy E is the sum of potential energy PE and kinetic energy KE. From $E = KE + PE$, $KE = E - PE$, and after substituting the latter into (1.16) and rearranging, one has

$$\frac{d^2\psi}{dx^2} + \frac{8\pi^2 m}{h^2}(E - PE)\psi = 0 \tag{1.17}$$

Further rearrangement of (1.17) yields

$$\frac{-h^2}{8\pi^2 m}\frac{d^2\psi}{dx^2} + PE\psi = E\psi \tag{1.18}$$

More simply, (1.18) can be written

$$H\psi = E\psi \tag{1.19}$$

where

$$H = \frac{-h^2}{8\pi^2 m}\nabla^2 + PE \tag{1.20}$$

∇^2 is the symbol for the second derivative d^2/dx^2 in one-dimensional x space. The term H is the operator equivalent for the electronic energy. Equation (1.17) is the Schrödinger equation in the one-dimensional x space (independent of time). *The first task of quantum chemistry is always finding ψ which solves this differential equation.* Once one has the molecular wavefunction ψ, one can compute molecular properties, as already stated. It should be noted that we knew the form of ψ first (1.12) and then the wave equation was set up. The problem is usually reversed, i.e., one knows all *but* ψ in Eq. (1.17).

The square of the wave function also has a meaning. Just as the square of the wave amplitude (height of the wave) is interpreted in electromagnetic radiation theory as a measure of intensity or photon density, ψ^2 in wave mechanics is a measure of density or probability. In Fig. 1.4, ψ^2 versus x represents a plot of the probability of finding the electron at various distances x.

For certain phenomena one must employ wave functions explicitly dependent on time (Eyring et al., 1944). In such cases a time-dependent Schrödinger equation must be solved, and the solution can be $\phi = \psi_x \psi_t$, i.e., ϕ will be time-dependent and a product of space and time functions.

Thus far we have shown that light may be treated as a wave train having certain values of energy. This wave may interact with matter, exerting force, and resulting in changes such as in the rotational or electronic energy of the molecule. From our point of view, the most important component of light interacting with electrons of molecules is the electric vector **E**.

The average position of a molecular electron may be described by using a molecular function. The form of the wave function describing the position of an electron in a molecule will depend on the energy of the electron. Thus, if light interacts with an electron so as to cause a transition, the molecule must be described by a new wave function. There are, however, only discrete energy levels which are allowed for each electron in a molecule. These may be calculated using the wave functions. The energy absorbed by the molecule must be the difference between the ground (lowest energy) state and the new state. This energy of the photon will be related to a certain wavelength, i.e., the wavelength of absorption.

When a molecule has absorbed the energy of a photon this increased energy results in a change in its structure. In the sections below we shall present several models which are used to describe such changes in the electronic structure of a molecule after the absorption of light. Two methods are used, the Morse potential energy curves and the molecular orbital model (linear combination of atomic orbitals).

C. MOLECULAR ENERGY CHANGES

Molecules may absorb light in the ultraviolet, visible, infrared, and microwave, etc., wavelength regions. These absorptions will cause electronic, vibrational, and rotational energy changes, respectively. In these absorption processes, molecules end up in excited states; they stay there for some time, may return to other lower energy excited states, and then return to the same or different ground states. Fields of spectroscopy can be categorized according to the type of changes (electronic, vibrational, etc.) occurring in the molecule or according to the energies required to bring about the changes. In Table 1.1 representative transition energies for three spectral regions are displayed in units of cm^{-1} rather than wavelength, since cm^{-1} is directly proportional to energy. The table shows order of magnitude estimates for electronic, vibrational, and rotational transitions, viz., 100, 5, and 0.01 kcal/mole.

TABLE 1.1 REPRESENTATIVE TRANSITION ENERGIES AND THE VALUE OF $\frac{1}{2}kT$ AT ROOM TEMPERATURE (300°K)

Spectral transition	eV	cm^{-1}	kcal/mole
Electronic	2.5	20,000	60
Vibrational	0.24	2,000	5.8
Rotational	0.0012	10	0.029
$\frac{1}{2}kT$	0.013	100	0.3

It should be noted that at 300°K the amount of thermal energy present can have an influence on the population of levels, i.e. it can be the cause of a thermal spectral transition. The thermal energy at this temperature is $\frac{1}{2}kT$ or 0.3 kcal. It is evident from Table 1.1 that this latter quantity of energy is greater than that required for a rotational transition to occur, but it is less than the energies needed for vibrational or electronic transitions to take place. Given an assembly of molecules the number which will be raised to higher energy levels may be calculated from the Maxwell–Boltzmann energy distribution equation (1.21). This equation expresses the relationship between population ratios of states, the energy difference between the states, the Boltzmann constant, and temperature:

$$\frac{n_s}{n_g} = e^{-\Delta e/kT} \qquad (1.21)$$

n_s/n_g is the ratio of numbers of molecules in excited [S] and ground [G] states and $\Delta E = E_s - E_g$, the energy difference between the states. Care should be taken that ΔE and kT are expressed in the same units (e.g., cm^{-1}, kcal).

One can compute a typical population ratio for ground and excited electronic energy levels at 300°K by employing Table 1.1 as follows:

$$\frac{n_s}{n_g} = e^{-60/0.6} = e^{-100} \cong \frac{1}{10^{37}} \tag{1.22}$$

Obviously, the thermal energy at 300°K cannot be responsible for getting molecules from ground into electronic excited states. Therefore, all molecules at room temperature are in their electronic ground states. This ratio (n_s/n_g) is much larger when rotational and vibrational energy changes are substituted into Eq. (1.21), and thermal energies at room temperature can appreciably populate these excited states.

The total energy E_t of a molecule in any given state can be considered the sum of three terms:

$$E_t = E_e + E_{\text{vib}} + E_{\text{rot}} \tag{1.23}$$

Vibrational transition ΔE_{vib} may be accompanied by rotational changes ΔE_{rot}, and vibrational transitions ΔE_{vib} may accompany electronic transitions ΔE_e. For example, fine structure due to $\Delta E_e + \Delta E_{\text{vib}}$ may appear on a band in the uv-visible region.

Empirical Morse potential energy curves of diatomic molecules (Wheatly, 1959) are useful to illustrate the relative positions of energy levels and transitions between them. Figure 1.5 shows two such curves for the electronic ground [G] and excited [S] states. Each point on the curve represents the net energy of the interaction types nuclear-nuclear, nuclear-electron, and electron-electron for that particular bond distance. v_i and v_i' are the ith vibrational levels in electronic ground and excited states, respectively. Furthermore, between any two vibrational levels v_k and v_{k+1} are several rotational energy levels which are not shown, since they do not concern us here. Curves of dashed lines represent vibrational probability distribution $|\psi_{\text{vib}}|^2$, the usefulness of which will be described below. ψ_{vib} symbolizes the wave function for the vibrational motion. These curves impart the following important additional information.

Curves S and G minimize at distances R_e^* and R_e, the equilibrium bond distances for excited and ground states, respectively. Whether R_e^* is shorter or longer than R_e depends upon the nature of the excited state (see Sec. 1.3). There is also a slight dependence of distance

R_e on the particular vibrational level the molecule is populating, viz., R_e for HBr is 1.420 Å and 1.423 Å in states $[G, v_o]$ and $[G, v_1]$, respectively (Herzberg, 1950). For a molecule in the ground electronic state $[G]$, vibrational transitions can occur between v-levels. Furthermore, it is of interest to note that the second derivative of the expression for $[G]$ or $[S]$ ground or excited electronic state with respect to interatomic distance R is the force constant, the magnitude of which may be considered proportional to the bond strength.

Figure 1.5 also shows that during the stretching motion of a molecule in state $[G, v_0]$ the atoms reach their maximum and minimum separations at R_{max} and R_{min}, respectively, but spend little time there ($|\psi|^2$ is small). In addition, the zero-point energy E_0, which is the vibrational energy of the molecule at absolute zero, for state $[G]$ is

$$E_0 = D_e - D_0 \qquad (1.24)$$

D_e and D_0 are considered negative quantities. The binding energy BE is another important molecular property which may be obtained from the curve. It is given by

$$\mathrm{BE} = -D_e \qquad (1.25)$$

Potential energy distribution of molecules with more than two atoms are much more complex.

1.3 Molecular Orbital (MO) Theory

A. THE ROLE OF MOLECULAR ORBITAL THEORY IN BONDING

We have seen that the excited and ground states of diatomic molecules were represented by two potential energy curves in Fig. 1.5. However, states of molecules having three or more atoms cannot be represented in this manner, since their multidimensional surfaces having many different minima are seldom known. It is informative, nevertheless, to examine the surface for a linear triatomic case.

The schematic potential energy surface for O=C=S is shown in Fig. 1.6. Curve XYZ in plane $ABCD$ (also see Fig. 1.7a) is the potential energy curves of bond O—C at the fixed C—S distance $R_{C-S} = A$. When R_{C-S} is given another value, say R_y, another cross section parallel to the first one must be taken (Fig. 1.7b). This new cross

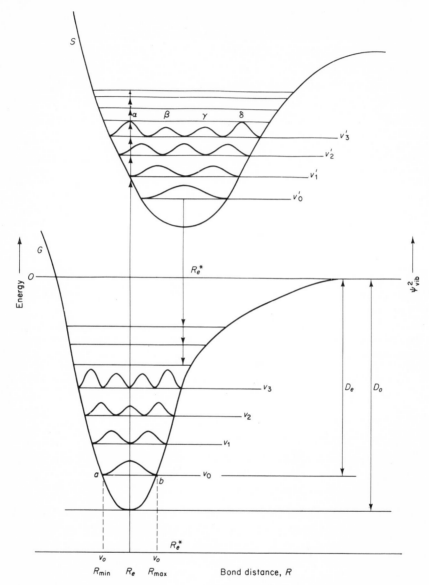

Fig. 1.5. Morse potential energy curves of ground $[G]$ and excited $[S]$ states for a diatomic molecule. (Schematic).

section will contain a new O—C potential curve, viz., the curve will be characterized by a new minimum R_{O-C} and new O—C force constant. The surface is also informative with respect to dissociation

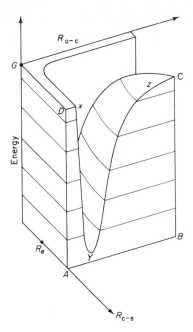

Fig. 1.6. Schematic potential energy surface for linear molecule O=C=S in one electronic state.

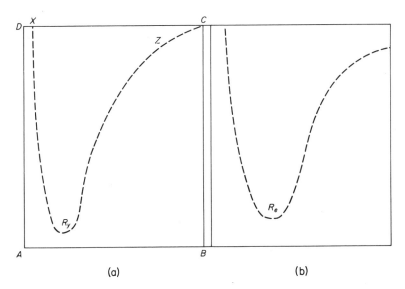

Fig. 1.7. Parallel cross sections in planes $ABCD$ (Fig. 1.6) (a), and at the distance R_e behind the same plane (b).

and "fusion." At coordinates and energy designated by point C, the molecule flies apart giving atoms O, C, and S. At point G the O, C, and S atoms would be "fused" together. While additional information is contained in Fig. 1.6, it should be emphasized that the complexity of such surfaces increases rapidly as the numbers of atoms and electrons in molecules increase. Surfaces for these triatomic and larger molecules are usually unknown. For this reason properties of large molecules are not investigated by this approach. However, it is possible to use molecular orbital theory for treating large molecules, and we shall give a brief account of this theory and indicate applications. For example, the theory permits one to assign the origin of many bands in electronic absorption spectra, and it allows one to understand electronic differences of molecules in ground and excited states.

B. CHEMICAL BONDS — APPLICATION OF THE MOLECULAR ORBITAL MODEL

We will give some simple examples of how we may apply this model. Before doing this, however, we have to recapture the fundamental concepts of chemical bonding theory. The general rules pertaining to the arrangements of electrons in atoms can be found in any basic textbook of chemistry, and we will only discuss those basic facts which are important for bond formation. Electronic orbitals in atoms (atomic orbitals) are classified according to their various orbital symmetries as s, p, d, etc., which are geometrically increasingly complex. For organic molecules, which are of primary interest in this text, we need consider only s and p orbitals. The simplest possible atomic orbital is an s orbital. For example, the electron of the hydrogen atom is in a $1s$ orbital (Fig. 1.8a). The "1" symbolizes the K shell of an atom. Other shells are $L = 2$, $M = 3$, etc. All s orbitals have identical symmetries; as indicated in the figure they are spherical, and the charge distribution is the same in all directions about the nucleus. In contrast, p orbitals have marked directional characters, they are dumbbell shaped, which means that they may be classified according to the directions of their long axes as p_x, p_y, and p_z orbitals, respectively. Furthermore, a p orbital possesses a nodal plane, i.e., a plane in which the probability of finding an electron is zero (Fig. 1.8b). The arrangements of electrons in orbitals in some elements of importance are given in Table 1.2, where the arrows within an orbital symbolize spin axes of the electrons. One orbital, may be occupied by one electron of any spin, but the second electron must be of opposite spin (Pauli's exclusion principle). Another rule, named after Hund,

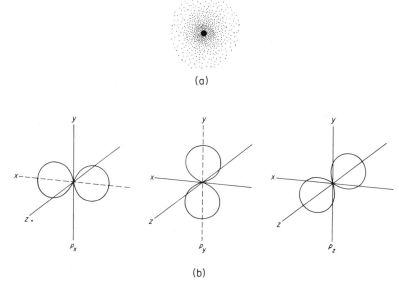

(a)

(b)

Fig. 1.8. (a) $1s$ orbital of hydrogen atom. (b) p orbitals.

TABLE 1.2 ATOMIC ORBITALS

		$1s$	$2s$	$2p_x$	$2p_y$	$2p_z$
1	H	(				
6	C	( )	( )	(	(	
7	N	( )	( )	(	(	(
8	O	( )	( )	( )	(	(

states that two electrons do not occupy the same p orbital until all p orbitals of that same shell have one electron each. It can be seen from the table that application of Hund's rule implies that carbon with two un- paired electrons in the $2p$ orbitals should be bivalent, which we know from organic chemistry is not so. The well-known tetravalent charac- ter of carbon is obtained by a process known as hybridization, which means, in the case of carbon, that a $2s$ electron is prompted into the empty $2p$ orbital, then the four orbitals mix to form four equivalent hybrid orbitals. The latter are called sp^3 hybrid orbitals (one s and three p orbitals).

The creation of most chemical bonds means the formation of a new orbital into which two electrons are fed, one from each of the two combining atoms. These two electrons will occupy the same new orbital, which means that they must have opposite spins (Hund's rule). We illustrate this with the simplest known molecule, the hydrogen molecule, in which both $1s$ electrons are fed into a new molecular orbital causing attraction between the atoms:

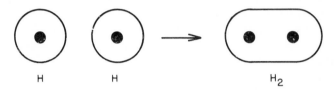

This orbital is symmetrical about the line connecting the two nuclei and hence has a symmetry similar to that of a $1s$ orbital. It is known as a σ molecular orbital. In general, the energy of a molecular orbital is low, and consequently the binding energy is high when the atomic orbitals overlap well. Two s orbitals have identical geometries and in H_2 they overlap very effiiciently. The covalent σ bond between the atoms in the hydrogen molecule is strong.

A somewhat more complicated molecule of the same general character as the hydrogen molecule is methane, CH_4, in which each of the four sp^3 hybrid orbitals of the carbon atom has combined with a $1s$ orbital of hydrogen atoms. Four bonding σ molecular orbitals form in a completely symmetrical fashion, i.e., the configuration is tetrahedral with valence angles of 109° 28'.

Generally, a σ molecular orbital can be formed from two s atomic orbitals, or from one s and one p atomic orbital, or from two p atomic orbitals which point toward each other. When both of the interacting atomic orbitals are p orbitals and overlap with each other in the sideway manner of Fig. 1.9a, the bonding orbital formed will be completely different. This is called a π orbital (Fig. 1.9b), and like the p atomic

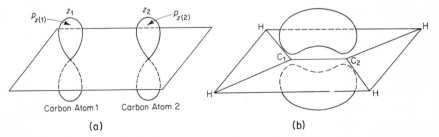

Fig. 1.9. Atomic orbitals p_z (a) that can participate to form the π-molecular orbital (b) in ethylene.

orbitals it has a nodal plane. The bond formed in this way is called a π bond, and the electrons involved are called π electrons. An example of a simple molecule possessing σ as well as π bonds is the ethylene molecule, $CH_2{=}CH_2$, where the bonds between carbon and hydrogen are σ bonds, whereas the carbon atoms are joined by one σ and one π bond (Fig. 1.9). Here carbon has undergone a different hybridization to form its bonds as compared to carbon in methane. This kind of σ hybridization in ethylene is called trigonal or sp^2 hybridization. Each carbon atom has four electrons available: a $2s$ pair and two electrons in p orbitals. π orbital formation occurs in connection with double or triple bond phenomena as in ethylene. In the so-called conjugated systems

$$-CH{=}CH{-}CH{=}CH{-}$$

we meet a particular kind of double bond. The unique properties of a conjugated system can only be explained by assuming that electronic changes at one end of the system result in a redistribution of charge throughout the system. The π electron at each atom in a conjugated system can be shared with both its neighbors so that the result is a many-centered delocalized orbital. In aromatic molecules we have the most complex systems of this kind. The symmetrical π orbitals of benzene, the typical aromatic molecule (Fig. 1.10), have a node in the plane of the molecule, which means that the probability of finding a

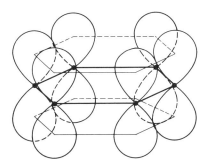

Fig. 1.10. Benzene. By interactions between the p orbitals indicated in the figure, doughnut-shaped π orbitals are formed above and below the plane of the ring.

π electron in the plane of the molecule is zero. In fact, the overlap of the π orbitals and the σ orbitals is zero, which means that the π electron system can be regarded as independent of the σ bonds.

Introducing two electrons with opposite spins into the same bonding molecular orbital is thus equivalent to forming a chemical bond. There is attraction between the atoms, although electrons may be fed into orbitals which cause repulsion between the atoms, i.e., the electrons may be engaged in antibonding. Orbitals may be classified as bonding, antibonding, or nonbonding. These characters may be indicated by giving the MO's the superscripts *b*, *, or *n*, respectively. Examples of nonbonding electrons important in absorption and fluorescence spectroscopy are the "free" electron pairs of oxygen (two pairs in case of bivalently bound oxygen) and of nitrogen (one pair in case of trivalently bound nitrogen).

The most common phenomenon associated with electronic excitation (absorption of light) is the introduction of electrons into antibonding orbitals. These electrons are often originally π electrons or nonbonding electrons, and the associated transitions are described as $\pi \rightarrow \pi^*$ and $n \rightarrow \pi^*$ transitions, respectively.

MO energy levels can be obtained as outlined in the preceding section. As an example, the MO energy level diagram of benzene is given below. The six p_z orbitals perpendicular to the plane of this molecule will form molecular orbitals $\phi_1, \phi_2, \ldots, \phi_6$, three bonding and three antibonding. The MO diagram of the π electrons is given in Fig. 1.11. Each pair of molecular orbitals ϕ_2, ϕ_3 and ϕ_4, ϕ_5 is of the same energy. ϕ_2 and ϕ_3 are called a degenerate set, as are ϕ_4 and ϕ_5.

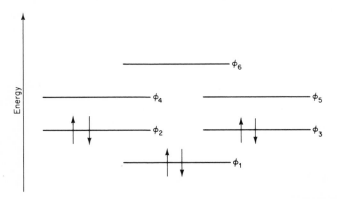

Fig. 1.11. The molecular orbital diagram of the π electrons of benzene.

In addition to giving an ordered picture of electronic structures, MO diagrams show the available unoccupied MO levels to which an electron may be excited. Selection rules, which formally tell if a selected transition is allowed, will not be treated here (see Daudel et al., 1959). In order to illustrate electronic excitation, let us consider the interesting and well-investigated example of formaldehyde. Figure 1.12 is a

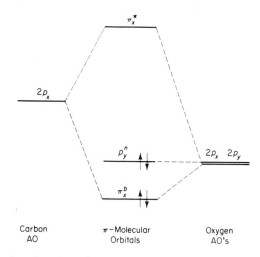

Fig. 1.12. Atomic and π molecular orbitals in formaldehyde (see Fig. 1.13).

partial MO diagram of the highest energy-filled molecular orbitals and the first empty π^* orbital. The molecule is in the y–z plane (Fig. 1.13).

The lowest energy transition will be from the nonbonding p^n molecular orbital to the antibonding π^* orbital which is indicated more briefly by $n \rightarrow \pi^*$ (Fig. 1.14). For example, the $n \rightarrow \pi^*$ transition in acetaldehyde is at 290 nm which is an indication of the separation between π^* and p^n levels.

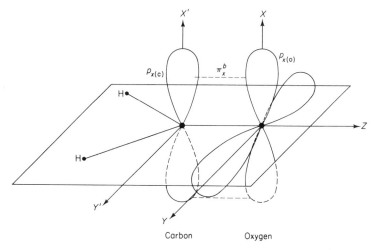

Fig. 1.13. Axis system in formaldehyde. The dashed line between p_x orbitals of carbon and oxygen represents the π bond they form (see Fig. 1.12).

The next highest transition is $\pi \rightarrow \pi^*$. This occurs at about 180 nm for saturated aldehydes. The entire analysis leads to the following important conclusion concerning the picture of the excited state. In this specific example, $n \rightarrow \pi^*$ or $\pi^b \rightarrow \pi^*$ electronic transitions are processes in which electrons are taken from MO's that are bonding (π^b) or have little or no effect (p^n) on the C=O functional group, respectively. However, the electrons end up in a π^*MO which weakens the binding between C and O in C=O (Fig. 1.13). Therefore, the ground and excited states of H_2C=O are expected to have significantly different electronic arrangements and the C=O distance is expected to be larger in the excited state. Such considerations as these also explain $R_e \neq R_e^*$ of Fig. 1.5. It is useful to keep in mind molecular orbital results such as these, since they relate to electronic transitions, bonding, and the nature of excited states and ground states in a very general way.

1.4 Electronic Transitions, Oscillator Strengths, and the Lambert–Beer Law

We adopt the notation $[G]$, $[S_1]$, $|S_2]$, and $[S_3]$ to symbolize the electronic ground state, first, second, and third, spin-singlet excited states, respectively (see Figs. 1.14 and 1.15). Whenever the excited state is written as $[S]$,* it is implied that we mean $[S_1]$ unless otherwise made clear. First, second, and third excited electronic spin–triplet states will be represented by $[T_1]$, $[T_2]$, and $[T_3]$, respectively.

The absorption (or emission) band height or strength of a spectral band can be described in several ways. The Lambert–Beer law is used to measure such absorption strengths experimentally, and is expressed by

$$-\log \frac{I}{I_0} = \epsilon c l \qquad (1.26)$$

*The nomenclature $[S]$ arises from the quantum mechanical state of the molecule. In organic molecules all of the electrons are paired in the ground state $[G]$. That is, for every electron with a spin of $+\frac{1}{2}$ there is one with a spin of $-\frac{1}{2}$. The most probable excited state also has all the electrons paired. The spin state of a molecule is defined as $[S] = [2\,(\text{spin of electron } 1 + \text{spin of electron } 2 + \cdots \text{electron } n)] + 1$. Since in the case of an organic molecule when all the electrons are paired this becomes $2(0) + 1 = 1$, the molecule is in a singlet state.

If the excited electron becomes unpaired there will be two unpaired electrons in the molecule and both will have the same sign, thus:

$$S = 2[\tfrac{1}{2}E_1 + \tfrac{1}{2}E_2 + \cdots] + 1$$
$$= 2[\tfrac{1}{2} + \tfrac{1}{2}] + 1 = 3$$

This molecule is in a triplet excited state. See Chap. 1, Sec. 4 for a more complete description.

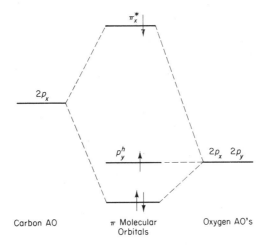

Fig. 1.14. Singlet excited state of formaldehyde. For triplet state, π_x^* electron would have opposite spin.

where ϵ is the molar extinction coefficient or molar absorbancy index (a molecular property), c the concentration of absorbing species in moles-liter^{-1}, and l is the path length in cm. Each band has its characteristic value of ϵ. I_0 and I are the light intensities as the light enters and leaves the absorbing medium (Fig. 1.16).

Since most spectrophotometers record absorbancy, $A = \log(I_0/I)$, also called optical density, the Lambert–Beer law is most commonly used as

$$A = \log\frac{I_0}{I} = \epsilon c l \qquad (1.27)$$

The reader is referred to Swinehart (1962) for a concise derivation of this law. It is often employed in form (1.27) to determine the unknown concentration of solute molecules that have a characteristic ϵ value at the wavelength of a specific absorption (or emission) band maximum. The failure of a substance to obey the Lambert–Beer law may be an indication that it is binding with another moiety present in the same solution.

The molar absorbancy index ϵ is characteristic of the absorbing molecule; this ϵ value is a direct measure of the ability to absorb light. It is expected that the absorption intensities are related to quantum mechanical absorption strength of the band. These strengths can be predicted by theory if the wave functions of the ground $[G]$ and excited $[S]$ states are known. The factors which must be considered and which determine whether or not photons will be absorbed by a

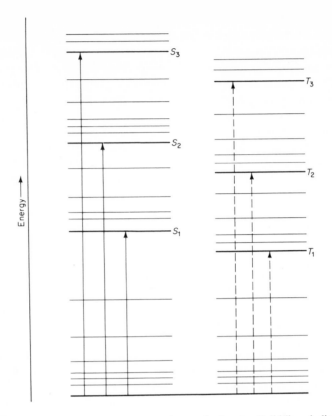

Fig. 1.15. Electronic transitions from ground to excited states. Solid lines indicate G–S transitions. Dashed lines indicate less probable G–T transitions.

molecule (and also for emission) are the populations of the two states $[G]$ and $[S]$, and the strength of the interaction between the electromagnetic radiation and the molecule. The first quantity understandably

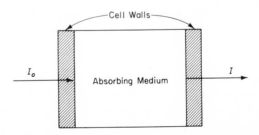

Fig. 1.16. Intensities of a monochromatic light beam entering and leaving the absorbing medium.

answers the question of whether or not the molecule can absorb the radiation. If a large proportion of molecules are in the excited state $[S]$ then there are fewer in the ground state which can absorb the radiation. The interaction magnitude is considerably more complicated and depends on which molecular orbital the electron in question occupies, the type of states $[S]$ and $[G]$, and the relation of the two states to one another.

If one considers all the possible mechanisms by which an electron in a molecule could undergo a transition these would include collision transfer, spontaneous absorption of light, induced absorption, spontaneous emission, and induced emission. Since it is very improbable for an electron to jump into a higher orbital in the absence of light, the second process will not be discussed [see Eq. (1.22)]. Induced absorption of light of frequency v results in the absorption of a photon. This places the molecule A into the excited state A^*. Conventionally the coefficient assigned to this process is designated as B with subscripts G and S designating the transition from ground to excited state. This coefficient may be considered a constant relating to the efficiency of the process. Thus

$$A + hv = A^*; B_{G,S} \tag{1.28}$$

The coefficient B contains the wave function for the electron in the ground state and in the excited state. It also has factors relating to the dipole nature of the electronic transition. One simple picture might be that of the probability of electric dipole interacting with an electric field.

Spontaneous emission from the excited state A^* to the ground state also has an associated coefficient relating to the efficiency of this process. This is designated by coefficient A with subscripts S and G.

$$A^* = A + hv; A_{S,G} \tag{1.29}$$

Induced emission, that is, the transition from A^* to A in the presence of an electromagentic field hv, will not be considered. For the case of interest here the amount of energy normally exciting a fluorescent sample will not produce a great enough population of molecules A^* so that the number of molecules in the ground state will change.

The terms $B_{G,S}$ and $A_{S,G}$ are referred to as the Einstein coefficients of induced absorption and spontaneous emission, respectively. $A_{S,G}$ and $B_{G,S}$ are related to one another by the equation

$$A_{S,G} = 8\pi hcv^3 B_{G,S} \tag{1.30}$$

where h is Planck's constant, c is the speed of light, and ν is $1/\lambda$ or cm^{-1}. Both $A_{S,G}$ and $B_{G,S}$ are independent of temperature. The induced absorption coefficient $B_{G,S}$ is a much more available quantity than $A_{S,G}$ by both quantum mechanical calculation and experimental evaluation. Equation (1.30) also shows that A and B are directly proportional to each other. In other words, molecules which absorb light readily will also emit it readily, those which absorb light unfavorably also have an unfavorable coefficient of emission. A corollary of this is that when emission is unfavorable the electron spends a long time in the excited state.

The molecular absorption of light may be described by considering the molecule as an oscillating dipole. A quantity referred to as the oscillator strength can be calculated and related to experimental extinction coefficient ϵ of the molecule. In fact, the *oscillator strength $\bar{f}$* is related to the integrated extinction coefficient. Equation (1.31) may be used to approximate the experimental oscillator strength of absorption bands:

$$\bar{f}_{G,S} = 4.32 \times 10^{-9}, \quad \int_a^b \epsilon(\nu) \, d\nu \qquad (1.31)$$

Since we are using a ratio of the observed to theoretical absorption band, values of $\bar{f}$ are normalized so that 1 is a maximum value. Molecules such as fluorescein with strong absorption bands approach 1 while those of weaker absorption bands such as 1-anilino-8-napthalene-sulfonic acid have $\bar{f}$ of about 0.07.

Since $B_{G,S}$ and $\bar{f}$ are measurements of the same quantity there is then a relationship between $B_{G,S}$ and the integrated extinction coefficient. This relationship can be shown to be

$$B_{G,S} = \frac{4.32 \times 10^{-9}(\pi)}{\nu m_e c^2} m \int_a^b \epsilon(\nu) \, d\nu \qquad (1.32)$$

The spontaneous coefficient of emission $A_{S,G}$, since it is related to $B_{G,S}$, is also related to the extinction coefficient. Thus knowledge of the ground state can yield important information with respect to the excited state and the lifetime of the excited state.

Before we progress to representative spectra, we will briefly explain the observation that molecular emission band maxima are observed at lower energies than absorption (excitation) band maxima (Fig. 1.5). Prior to irradiating the molecules, they are assumed to be in the ground vibrational state of the ground electronic state, or $[G, v_0]$. A photon of appropriate energy can excite molecules to state $[S, v_2']$. This energy is

$$\Delta E_a = E[S, v_2'] - E[G, v_0] \qquad (1.33)$$

The molecule will quickly undergo vibration transition $v_2' \rightarrow v_0'$, and electronic emission will now occur to state $[G, v_4]$. The energy of the emitted photon (fluorescence or phosphorescence) will be

$$\Delta E_e = D[S, v_0'] - E[G, v_4] \qquad (1.34)$$

It is evident from Fig. 1.5 that emission energy ΔE_e is less than absorption energy ΔE_a. In fact, the *difference* in energy of absorption and emission maxima, SE, is approximately $\Delta E_a - \Delta E_e$.

1.5 Spectra

A. Representative Spectra, Their Interpretations, and Applications of Principles

Biochemically interesting molecules, i.e., organic molecules, usually have one or more useful electronic absorption bands between 200 mμ and 800 mμ. A "useful" absorption band is one which may be employed for routine assay work in conjunction with the Lambert–Beer law [Eq. (1.37)], or as an excitation band for fluorescence. For example, benzene, naphthalene, and anthracene have absorption bands at 255, 315, and 380 mμ, respectively.

It is convenient to speak of bands in the visible, near uv, medium uv, and far uv regions of spectra. These are between 800 and 400 nm, 400 to 300 nm, 300 to 200 nm, and below 200 nm, in the same order.

Unsaturated carbon compounds are those containing π electrons by definition. A group of atoms containing π electrons is a chromophore. For purposes of interpreting spectra one makes use of the fact that a band system can often be related to a specific chromophore (e.g., $>C{=}O$, $>C{=}C$, $>C{=}S$). For example, the $C{=}O$ chromophore in acetone is responsible for a band at 290 nm.

It may be generalized that uv-visible spectra of organic molecules partially or entirely involve π electrons. For example, when an electron jumps from a bonding orbital to a antibonding orbital, the transition labeled $\pi \rightarrow \pi^*$, essentially only electrons are affected. Transition types $n \rightarrow \pi^*$ also give rise to band systems, and arise when an electron

localized on one atom (e.g., nitrogen in $\left(\,:\,\right)$ N⬡ or pyridine) jumps

to a vacant π antibonding orbital. Details of these electronic transitions are described below.

B. Nomenclature and Spectra

Spectra of aromatic molecules have several features in common, and a general elucidation can be achieved by describing the benzene spectrum. Figure 1.17 displays the first three band systems which are at 255 nm, 205 nm, and 183 nm. These band systems arise from $\pi \rightarrow \pi^*$ transitions, and the excited states have been labeled in several ways due to lack of general agreement by chemists to employ a universal system of nomenclature. Three equivalent sets of labels are shown in Fig. 1.17 and Table 1.3.

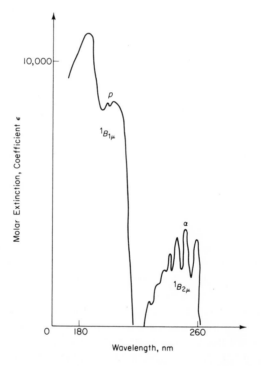

Fig. 1.17. The electronic absorption spectrum of benzene.

Group theory labels for states are universally understood by spectroscopists and are employed in this text. The ground state of benzene is $^1A_{1g}$, and it can be shown that all three excited states $^1E_{1\mu}$, $^1B_{1\mu}$, and $^1B_{2\mu}$ are spin-allowed (ground and excited state have the same spin state, $S = 0$). The spin is implied in the superscript, e.g., for state $^1B_{2\mu}$, $2S + 1 = 1$, therefore $S = 0$. Condition $S = 0$ means all electronic spins are paired.

TABLE 1.3 NOMENCLATURE OF BAND SYSTEMS IN BENZENE

	Molar extinction coefficients and location of bands[a]		
Nomenclature	183 nm	205 nm	255 nm
Group theory	$^1E_{1\mu}$	$^1B_{1\mu}$	$^1B_{2\mu}$
Platt	$^1B_a, {}^1B_b$	^1L_a	^1L_b
Clar	β	p	α
ϵ_{max}	100,000	10,000	100

[a]Labels B, E, β, etc., identify the excited state. It is generally understood what the ground state of a molecule is.

Another way of looking at the spin-allowed transitions is to state that generally it is very improbable for an excited electron to change its spin, i.e., an electron with a spin of $+\frac{1}{2}$ in the ground state will very probably have a spin of $+\frac{1}{2}$ in the excited state. The transition is "allowed," viz., of very high probability. In this case, it is spin allowed. See Fig. 1.14 for one electron configuration of a spin-allowed transition of formaldehyde. If an excited state required one electron to change its spin to $-\frac{1}{2}$ ($2S + 1 = 3$ or triplet state) the transition would not be spin allowed or it would be a "forbidden" (very improbable) transition.

The symmetry of the ground state $[G]$ is often quite evident from other experimental data, so that bands are identified by designating the excited states. From the spectrum of benzene it is evident that $^1B_{1\mu}$ and $^1B_{2\mu}$ bands are much weaker than the $^1E_{1\mu}$ band. The reason for this difference is that the first two are not allowed by group theory to absorb dipolar radiation. However, certain vibrations can distort the molecule and make the transition allowed.

The presence of the fine structure on, for example, band $^1B_{2\mu}$ should be noted. It is the vibrational–electronic fine structure mentioned above. This means that the spikes correspond to transition such as

$$[G, v_i] \rightarrow [S, v'_1]$$
$$\rightarrow [S, v'_2]$$
$$\cdot$$
$$\cdot$$
$$\cdot$$

In other words, both electronic and vibrational transitions occur. The spikes are separated by $\Delta E = v'_2 - v'_1$, etc. This is presented pictorially by Fig. 1.18 which shows that the energy separations of spikes

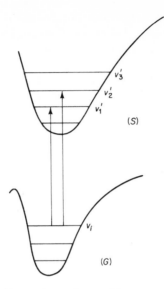

Fig. 1.18. Vibronic transitions: electronic transitions are accompanied by vibrational transitions.

correspond to the energies necessary for exciting molecular vibrational modes in the electronic excited state $[S]$.

These three band systems exhibited by benzene appear repeatedly in spectra of other aromatic molecules, i.e., the spectrum of napthalene appearing in Fig. 1.19. It is evident from this figure that the general

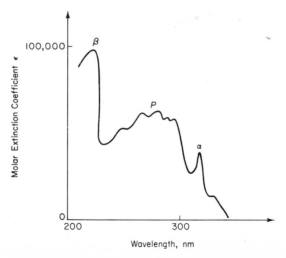

Fig. 1.19. Electronic absorption spectrum of naphthalene.

features of the benzene spectrum are present. The excited states, however, have different labels because benzene and naphthalene have different geometries (the electronic arrangements or states are, of course, also different). The naphthalene bands (excited states $^1B_{1\mu}$, $^1B_{2\mu}$, and $^1B_{3\mu}$ also labeled α, p, and β) are at 315 nm, 290 nm, and 221 nm. They arise from $\pi \rightarrow \pi^*$ transitions as did the benzene bands. (Compare Fig. 1.19 to Fig. 1.17.)

Mention is made of molecular spectra of aromatic compounds containing heteroatoms and of heteroatoms which are π-bonded to the aromatic ring. These systems have spectra similar to parent compounds, e.g., compounds containing only carbon and hydrogen. In addition to $\pi \rightarrow \pi^*$ transitions, $n \rightarrow \pi^*$ transitions can also be observed. During $n \rightarrow \pi^*$ transitions an electron is excited from a nonbonding orbital n localized at an atom to a π antibonding orbital π^* which can extend over all atoms in the molecule.

Both aniline and pyridine contain a nitrogen heteroatom. Figure 1.20 shows the n-orbital of pyridine which participates in the $n \rightarrow \pi^*$ transition. The same figure also shows that aniline cannot have an

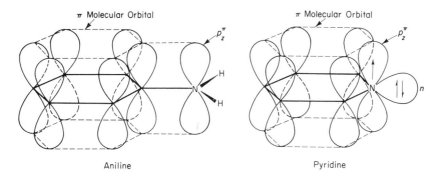

Aniline   Pyridine

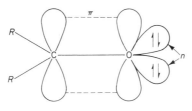

Carbonyl Chromophore

Fig. 1.20. Electrons at the nitrogen heteroatoms in aniline and pyridine, and at the oxygen in carbonyl compounds.

$n \rightarrow \pi^*$ transition because all electrons are employed for bonding to other atoms, i.e., the nitrogen electrons participate in π-bonding. Figure 1.21 contains the pyridine spectrum and molecular orbitals. Table 1.4 contains the information for labeling $n \rightarrow \pi^*$ transitions and the expected magnitude of molar extinction coefficients.

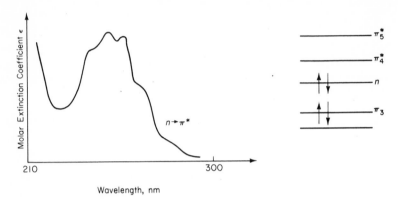

Fig. 1.21. Absorption spectrum and π-molecular orbitals in pyridine.

TABLE 1.4 NOMENCLATURE FOR $n \rightarrow \pi$ EXCITED TRANSITIONS AND THEIR EXTINCTION COEFFICIENTS

Information	Transition	
Platt nomenclature	$^1W \leftarrow {}^1A$	$^1U \leftarrow {}^1A$
ϵ_{max}	100–1000	10–100
Increasing solvent polarity	Blue shift	Blue shift
Examples	Pyridine	Acetone

Acetone has its $n \rightarrow \pi^*$ band at 290 nm. This is a singlet–singlet $(G \rightarrow S)$ transition. It is formally forbidden by dipole selection rules and consequently has very weak intensity. The spin-forbidden singlet–triplet $(G \rightarrow T)$ $n \rightarrow \pi^*$ transition occurs at 400 nm. It is also very weak ($\epsilon \cong 10^{-3}$) because *most* singlet–triplet transitions are forbidden. It is known that $n \rightarrow \pi^*$ fluorescence (emission $\pi^* \rightarrow n$ is actually meant) is generally not observed because of radiationless deactivation to the triplet level T (Fig. 1.22).

An important property of $n \rightarrow \pi^*$ bands is that increasing the solvent polarity shifts the bands toward shorter wavelength (blue shift) (see Chapter 3). This has been attributed to the interaction of electrons in nonbonding orbitals n with the solvents. In other words, increasing the

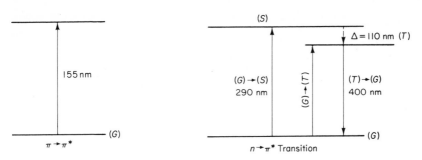

Fig. 1.22. $\pi \to \pi^*$ absorption and the $n \to \pi^*$ absorption and emission processes of the carbonyl chromophore.

polarity decreases the energy of the ground state, which corresponds to a more stable state. In effect this causes a greater separation between ground and excited states. The excited state is assumed to be much less effected by the solvent variation because the electron is no longer in orbital n, therefore contributing less to the molecular polarity (Fig. 1.23).

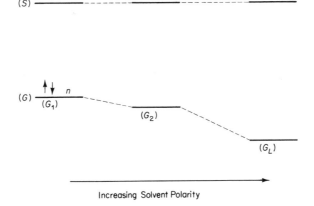

Increasing Solvent Polarity

Fig. 1.23. Effect of varying solvent polarity on the energy of the ground state (molecules having nonbonding electron).

The extreme interaction of nonbonding electrons with a positive charge is the protonation of the site. The pyridinium ion is formed in this way, for example:

The $n \rightarrow \pi^*$ band will be absent in the electronic spectrum of (**2**), because the original *n*-electrons now constitute the bond between the nitrogen and H^+.

Only the absorption band of lower energy is of interest in fluorescence spectroscopy with few exceptions. Energy at this wavelength serves to put molecules into excited states $[S_1]$ from which they fluoresce (or phosphoresce) in the process of returning to the ground electronic state $[G]$. For example, fluorescence bands of tyrosine, tryptophane, and phenylalanine are at 305 nm, 340 nm, and 280 nm, respectively. They are obtained by irradiating these molecules at the wavelengths of absorption band maximia at 275 nm, 280 nm, and 250 nm, in the same order.

The fluorescence and absorption bands of anthracene are shown in Fig. 1.24. The fluorescence band results from irradiating a solution of anthracene in the wavelength region of its absorption band maximum, 380 nm.

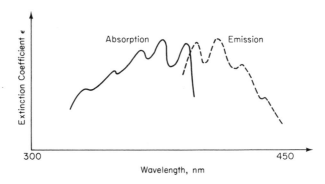

Fig. 1.24. Absorption and fluorescence bands of anthracene.

REFERENCES

Daudel, R., Lefebvre, R. and Moser, C. (1959). *Quantum Chemistry*, Wiley-Inter-science, New York.
Davisson, C., and Germer, L. (1927). *Phys. Rev.*, **30**, 705.
de Broglie, L. (1925). *Ann. Physiol.*, **3**, 22.
Eyring, H., Walter, J., and Kimball, G. E. (1944). *Quantum Chemistry*, Wiley, New York, p. 107.
Glasstone, S. G. (1946). *Textbook of Physical Chemistry*, Van Nostrand, Princeton, New Jersey.
Herzberg, G. (1950). *Spectra of Diatomic Molecules*, Van Nostrand, Princeton, New Jersey.
Roberts, J. D. (1961). *Notes on Molecular Orbital Calculations*, Benjamin, New York.

Simple examples of applying an approximate version of molecular orbital theory for organic molecules may be found in the book by J. D. Roberts,

Swinehart, D. F. (1962). *J. Chem. Ed.*, **39**, 333.
Wheatley, P. J. (1959). *The Determination of Molecular Structure*, Oxford Univ. Press, London and New York.

Chapter 2

Fluorescence of Absorbed Radiation

2.1 Absorption and Fluorescence

A. ABSORPTION AND FLUORESCENCE PROCESSES

When light interacts with matter, the photons may either collide or be absorbed. If the collisions are elastic (the impinging and dispersed radiation are of the same wavelength and are radiative in a somewhat random manner like a billiard ball rebounding from an object) it is called Rayleigh scattering. The extent of scattering by this process is wavelength dependent and follows ideally the relationship of $1/\lambda^4$. If the collisions are nonelastic, due to mixing the electromagnetic energy with the rotational and vibrational energy of the colliding molecule, the emerging radiation will be of a different wavelength. In this case, if the absorbing moiety were in the ground state (electronic and vibrational) previous to the interaction, the emerging photon will be of less energy. This nonelastic collision is less probable in occurrence than the elastic one and is referred to as Raman scattering (Fig. 2.1).

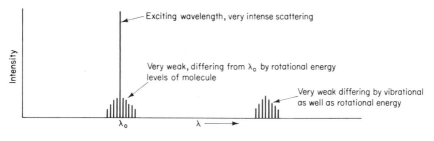

Fig. 2.1

37

Several quantum mechanical considerations are important in the absorption and loss of interacting radiation: (1) there are discrete steps between one electronic, vibrational, and rotational level to another, although in the condensed phase most systems will give continuous spectra due to secondary interactions; (2) if the absorbed photon has more energy than needed for a simple electronic transition, the excess energy is usually absorbed as vibrational and rotational energy.

In Fig. 2.2 the transition from ground to excited state is shown by an arrow. *The change from ground state to excited state is essentially instantaneous compared to the time necessary for the nuclear coordinates to change.* This rule is referred to as the *Franck–Condon* principle and is discussed in more detail in Chapter 3 (Condon and Morse, 1929).

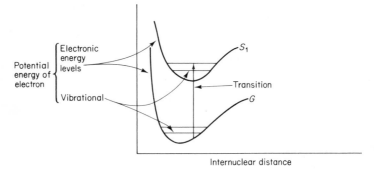

Fig. 2.2

For a molecule excited by light the transition to the excited state is very rapid, of the order of 10^{-15} sec. Transitions to the lowest singlet state are less rapid ($\sim 10^{-11}$ sec), and finally the probability of the transition to the ground state is even slower ($\sim 10^{-9}$ sec). The time scale for changes in the nuclear coordinates are on the order of 10^{-12} sec.

There are many possible electronic energy levels for a particular electron. If light is absorbed by the absorption band of least energy (the longest wavelength band), then the molecule will probably undergo a transition from the ground state to the next highest electronic energy level $[S_1]$. If light with more energy is absorbed by the molecule it may attain an even higher electronic energy level $[S_2]$ or $[S_3]$. If this does occur, this excess energy is quickly lost and the molecule usually progresses to the lowest excited state $[S_1]$ (Fig. 2.3).

The type of transition discussed here is the absorption of light resulting in the conversion of the molecule into a higher energy state.

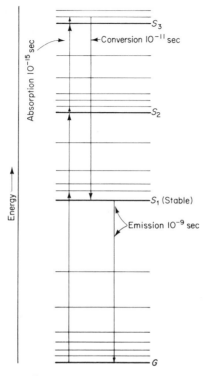

Lifetimes of electronic transitions

Fig. 2.3. Lifetimes of electronic transitions.

There are, however, several types of chemical reactions which will put the molecule in a similar high energy state. In some of these reactions the molecule can revert to the ground state by emitting a photon. The amount of energy necessary for the molecule to be in such an excited state is so great that these reactions are rare and many of the most commonly studied involve molecular oxygen.

When the impinging photon is absorbed by an atom or molecule, it returns to the ground state by one of several paths. The excited molecule may convert all of the excitation energy into internal vibrational and rotational energy. This may be dissipated as heat to solvent molecules if in solution, or as a change in molecular structure such as ionization or molecular dissociation. Such transitions are termed radiationless since no photon is emitted in the visible or uv regions. The next set of pathways involves passage through the lowest singlet excited state. In most atoms or molecules this has a much greater probability than any of the other excited states. The excited molecule can convert a part of the absorbed energy into vibrational and rotational energy, which is eventually lost to solvent molecules, and can then exist for some time in the lowest excited state $[S_1]$. Loss of the absorbed energy from this metastable state to the ground state may proceed by a variety of pathways. The energy may be transitioned into internal vibrational and rotational energy as described above, or the excited molecule may interact and transfer the energy to a colliding or complexed molecule (see Sec. 2.4). For molecules in solution this may be a solvent molecule (see Sec. 3.2). However, there are molecules such as I^- which are very efficient absorbers of this type of energy. The energy may be dissipated as a photon and return directly to the ground state. This is what we call fluorescence. There is also a probability that the excited electron will change its spin. The molecule is then said to be in a triplet excited state. Normally this conversion has low probability. Transition from the triplet state to the ground state may occur by mechanisms similar to those described for descent from the singlet to the ground state: internal energy conversion, collisional transfer, and emission of a photon (phosphorescence). The excitation energy from both singlet and triplet states may be transferred over a large distance (10 to 50 Å) to another molecule by a process referred to as resonance energy transfer (see Chapter 6).

Using the energy diagram again the deactivation pathways of a molecule can be illustrated as shown in Fig. 2.4.

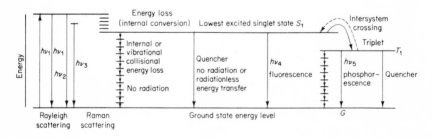

Fig. 2.4

Where the energy is lost by transition through various vibrational levels and energy is absorbed by the surrounding solvent, the energy diagram shows that the two electronic levels could be so oriented that a small energy loss could result in a change in electronic states. In Fig. 2.5a, this would be the small difference depicted by the arrow.

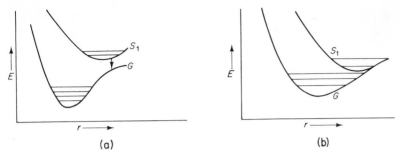

Fig. 2.5. Energy diagram where r is in the internuclear distance and E the energy.

Radiation would occur in the infrared region of the spectrum, and the state of the electron would change from $[S_1]$ to $[G]$. If the excited state overlaps the ground state (Fig. 2.5b), the molecule may easily transform from one of state $[S_1]$ to one of ground state electronic energy but with large amounts of vibrational energy.

The mechanism of interaction of a molecule in the excited state with one which absorbs the excitation energy efficiently upon collision, but does not fluoresce, is not completely understood. The molecule that absorbs the energy is called a quencher, and the term Q is normally used for abbreviation:

$$[\text{Excited molecule}]^* + Q \rightarrow [\text{Ground state molecule}] + Q^*$$
$$\downarrow \text{no photon but heat}$$
$$Q \qquad (2.1)$$

The energy of the quencher ultimately goes to the solvent, most probably through a series of very fast vibrational changes. A more complete description of this process is developed in Sec. 2.4.

B. RELATIONSHIP OF ABSORPTION AND EMISSION SPECTRA

The probable electronic and vibrational transitions for a molecule from the ground state to the lowest excited state are illustrated in Fig. 2.4 (Barrow, p. 232). This corresponds to the absorption of a photon of energy by the longest wavelength absorption band. The two energy wells are displaced from one another, reflecting the difference in the electronic structure of these states. Normally the excited state $[S_1]$ is broader in the nuclear ordinate direction and displaced toward larger internuclear distance than the ground state. The horizontal lines $\nu = 1, 2$, etc., or $\nu^1 = 0, 1, 2$, etc., define the vibrational levels of the molecule. The shaded areas reflect the probability of where the

electron would be if it were in that vibrational level. Transitions from
[G] to [S₁] depend on the probability of where the electron is in the
ground state and where it would be if the transition did occur. Since
the Franck–Condon principle states that there is virtually no change in
the nuclear coordinates during the transition, vertical lines may be
drawn describing the transition. (See Chapter 3 for a description of this
principle.) Those transitions which have the highest probability will
describe the excitation energy which will most probably be absorbed.

A simple illustration of this is to place a straight edge on Fig. 2.6
parallel to the energy axis. The most favored transitions will be those

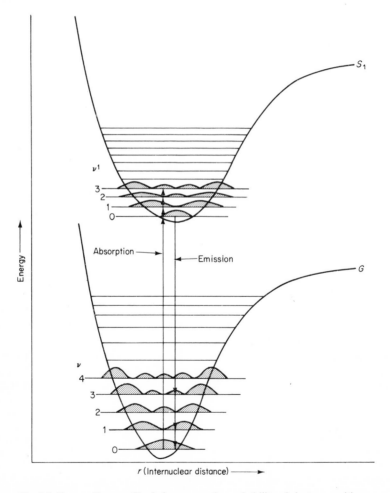

Fig. 2.6. Energy diagram. Shaded areas are the probability of electron position.

in which the shaded areas of both the ground and excited states are a maximum. In evaluating the transitions from $\nu = 0 \rightarrow \nu^1 = 0,1,2$, etc., or from $\nu = 0,1,2 \rightarrow \nu^1 = 1$, etc., the probabilities of all these transitions may be calculated. In practical limits the most probable state for an electron at room temperature is at $\nu = 0$ in the ground state, or at $\nu = 1$, or even at $\nu = 2$ if there is little difference in energy (600 cal mole^{-1})* between $\nu = 0$ and $\nu = 2$. This is seen clearly from the absorption spectrum of molecules. A molecule with a close spacing of vibrational states at its lowest energy absorption band would have a spectrum such as Fig. 2.7.

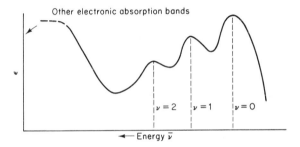

Fig. 2.7. Diagram of energy versus extinction coefficient or absorbtivity (ϵ).

Presuming that the molecule has absorbed a photon from any part of the absorption spectrum of Fig. 2.7, the excited molecule has a variety of mechanisms which can dissipate the energy. Assume that it will deactivate by the emission of a photon. The molecule will dissipate some of the absorbed energy so that it is in the lowest energy singlet state. This means that the vibrational energy of the electron may be considered to be $\nu^1 = 0$ (Fig. 2.6).

The energy of the photon emitted will be the difference between the $\nu^1 = 0$ energy state and to whichever ground state the electron reverts. Again using a straight edge on Fig. 2.6 one can estimate the probability of returning to each vibrational state. Normally this is similar to the probability of the position of the electron in the ground state. The most likely transition would be from $\nu^1 = 0 \rightarrow \nu = 0$ followed by $\nu^1 = 0 \rightarrow \nu = 1$ and $\nu^1 = 0 \rightarrow \nu = 2$ (Fig. 2.8).

---

*The figure of 600 cal mole^{-1} is an approximation of the Boltzmann energy distribution between the molecules at room temperature (300°K). The calculation is $E = NkT$ where k is the Boltzmann constant per molecule and N is Avagadro's number.

$$E = 6 \times 10^{23} \times 1.37 \times 10^{-16} \text{ erg/deg} \cdot 300 \text{ deg}$$
$$E \sim 600 \text{ cal mole}^{-1}$$

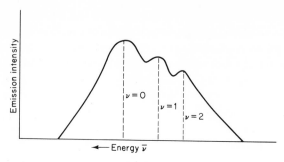

Fig. 2.8

Now if both the absorption and emission spectra are put together the fluorescence spectrum is a mirror image of the absorption spectrum (Levshin, 1931; Birks and Dyson, 1963) (Fig. 2.9). Experimentally this is observed for many aromatic compounds. However, this symmetry is dependent on the position and shape of both the ground state and excited state energy levels. Molecules having excited states significantly different from those given here will not have mirror image symmetry (Fig. 2.6). If the excited state $[S_1]$ trough were removed toward the right (greater distance) there would be less correspondence between the vibrational modes of the $[G]$ to $[S_1]$ transitions and the $[S_1]$ to $[G]$ transitions. The most favored $[G]$ to $[S_1]$ transitions might, for example, be $\nu = 0 \rightarrow \nu^1 = 5$ while the most likely $[S_1]$ to $[G]$ transition might be $\nu^1 = 0$ to $\nu = 3$.

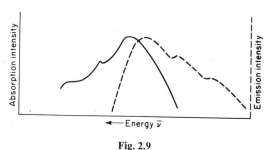

Fig. 2.9

2.2 Quantum Yield

A. Quantum Yield, Conversion Processes, and Lifetime of the Excited State

When a molecule is excited there are a variety of processes which will return it to the ground state. These are briefly described in Chapter 1.

If we consider three processes for returning to the ground state (radiationless energy loss, intersystem crossing through the triplet state, and emission of a photon) then the efficiency of emission will be a function of the competing rates of these processes (Hercules, 1966):

$$q = \frac{k_f}{k_f + k_i + k_x} \qquad (2.2)$$

where q is the efficiency or quantum yield and k_f is the rate constant for fluorescence emission, k_i the rate constant for radiationless energy loss, and k_x the rate constant for intersystem crossing. The term k_f also relates to the average lifetime of the excited state τ_0 by the equation $k_f = 1/\tau_0$. The average lifetime is used since any one molecule can emit light at many probable times, much less or much greater than τ_0.

The quantum yield is simply the ratio of the number of quanta absorbed. It is a dimensionless quantity:

$$q = \frac{\text{number of photons emitted}}{\text{number of photons absorbed}} \qquad (2.3)$$

From the discussions in the previous chapter, if there were no other factors involved the fluorescence emission and absorption should be roughly proportional to one another. For example, if only the $\nu = 0$ to $\nu^1 = 0$ and $\nu^1 = 0$ to $\nu = 0$ transitions are considered, the probability of absorption and that of emission should be identical. Therefore the extinction coefficient should be a simple function of the lifetime of the excited state. Molecules with high extinction coefficients should have a high fluorescence efficiency and a short lifetime of the excited state, i.e., k_f should be very large. Approximations used to calculate lifetimes from absorption spectra will be discussed below.

The factors that influence the state of radiationless energy loss are not completely understood but certainly are dependent on both the temperature and nature of the solvent. Finally, the amount of intersystem crossing to the triplet state is dependent upon the amount of triplet character of the singlet state as well as on the energy difference between the singlet and triplet states. In general, the greater the energy difference the smaller the rate constant.

B. MEASUREMENT OF τ

Direct measurement of the lifetime is based on the assumption that this process obeys the equation for first-order kinetics:

$$-\frac{d[A^*]}{dt} = k_f[A^*] \qquad (2.4)$$

where $[A^*]$ is the number of excited molecules. When the equation is integrated with respect to time it becomes

$$[A^*] = [A_0^*]e^{-t/\tau_0} \qquad (2.5)$$

where A^* is defined above, τ_0 is the lifetime $(1/k_f)$, t is time, A_0^* is the number of molecules in the excited state at $t = 0$ (see Fig. 2.10).

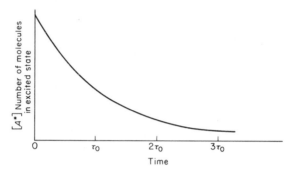

Fig. 2.10

If it were possible to excite a number of molecules instantaneously and measure the fluorescence as a function of time, then $t/\tau = 1$ when $[A_0^*]/e = [A^*]$ defines the lifetime of the excited state. This and other methods of determining the lifetime by direct measurement are described in Chapter 5.

Spectroscopic estimates of the lifetime of the excited state τ_0 or − as it may be inversely expressed − the rate of the transition from the upper (excited) state to the lower (ground) state may be calculated by methods based on two different principles, one based on concepts formulated on the absorption and emission probabilities and the other formulated from the uncertainty principle.

In the first approximation only three electric dipole transitions are considered:

(1) stimulated absorption
(2) stimulated emission
(3) spontaneous emission

The spontaneous absorption of energy which is equivalent to excitation of a molecule in the absence of an electric field is very improbable and consequently the most important mechanism to place a molecule in an upper state is by stimulated absorption. Similarly the emission of a photon from a dipole will be influenced if it is in a radiation field.

However, this transition may be neglected. In terms of experimental quantities this would mean that no external radiation fields would be applied and that the light or radiation stimulating the dipoles has too low an intensity to stimulate a dipole a second time (dipole–dipole interactions are described in Chapter 6). The most important process for returning to the lower state is then the spontaneous emission of a photon.

If now two further simplifying assumptions are made;

(1) the absorption band is very sharp
(2) the emission wavelength is the same as the absorption wavelength (see Fig. 2.11)

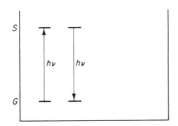

Fig. 2.11

(meaning that these conditions are applicable only to atomic transitions), then the probability of emission is directly related to the probability of absorption. The probability of emission is $1/\tau_0$, and the probability of absorption is the integral of the absorption over the transition in question ($\int \epsilon \, d\bar{\nu}$). The exact equation becomes

$$1/\tau_0 = 8 \times 2303\pi c \bar{\nu}e^2\eta^2N^{-1}\frac{g_e}{q_\mu} \int \epsilon \, d\bar{\nu} \qquad (2.6)$$

where c is the speed of light in a vacuum, $\bar{\nu}e$ is the emission frequency in cm^{-1} (which in atomic systems would be equal to the absorption frequency), η is the refractive index of the medium surrounding the oscillator, N is Avagadro's number, ϵ is the molar extinction coefficient of the last absorption band, and g_e and g_μ are the degeneracies of the lower and upper states, respectively (Förster, 1951). If the transition is singlet–singlet and the upper singlet has no degeneracies, g_e/g_μ would be 1. If the transition were singlet–triplet the value would be 3.

The most difficult quantity to estimate in Eq. (2.6) is ϵ because of overlapping of other transitions on the short wavelength side. In the worst case two electronic transitions may actually make up one band. The calculation may be simplified if one assumed the band to be a Gaussian distribution, where only the absorption maximum and half band width are needed in the calculation:

$$\int \epsilon \bar{\nu} \, d\gamma = \sigma \epsilon_m \pi^{1/2} \qquad (2.7)$$

where σ is the half band width and ϵ_m is the extinction coefficient at the maximum. For the dye 1-anilinonapthalene-8-sulfonic acid (Weber and Young, 1964), values for these quantities are

$$\eta^2 = 2.2, \quad \bar{\nu}e = 26{,}450 \text{ cm}^{-1}, \quad \sigma = 1900 \text{ cm}^{-1}, \quad \epsilon_m = 4.9 \times 10^3 \text{ M cm}^{-1}$$

Inserting these values into Eq. (2.4),

$$\frac{1}{\tau_0} = 2.88 \times 10^{-9} (2.2) \ (26{,}500)^2 \ (1900) \ (4.9 \times 10^3) \ (\pi^{1/2})$$

$$\tau_0 = 11.3 \times 10^{-9} \text{ sec}$$

This, however, is the calculation if the probability of emission and absorption were equal. If the measured quantum yield were less than 1, the experimental lifetime (τ_e) may be obtained by multiplying the theoretical lifetime by the quantum yield

$$\tau_e = \tau_0 q \qquad (2.8)$$

Berlman (1965) has devoted considerable effort to calculate and correlate a large number of the natural lifetimes with their experimental values. Another method for estimating lifetimes is based on the uncertainty principle but only gives a lower limit (Seybold and Gouterman, 1965).

The lifetimes may also be calculated from quenching experiments. For more detailed calculations see Sec. 2.4. Values of lifetimes may also be calculated from depolarization experiments (Chapter 4). For a further discussion of this topic, see Sec. 5.8.

2.3 Constancy of Quantum Yield

In condensed systems when molecules are excited the rate of change of this energy from upper electronic and/or vibrational states to the lowest singlet excited state are *very* fast compared to the rate of conversion of these excited molecules to the ground state.

Figure 2.12 is an idealized example of rates of conversion of energy from upper states to ground levels. These times are arbitrary limits, but of the correct order of magnitude for most molecular systems.

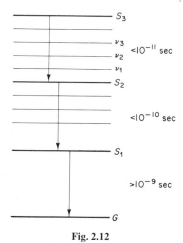

Fig. 2.12

For small molecules with well-integrated electronic systems, the competitive reactions from the conversion of excited molecules to the ground state by fluorescent or nonfluorescent processes (2.1) are *from the lowest excited state no matter how the molecules were excited.* Since all of the excited molecules must pass through this state, then regardless of which electrons (or which absorption bands) were excited, this lowest excited state will be formed. The efficiency of fluorescence will also be the same no matter how the molecules are excited. The quantum yield of fluorescence is therefore independent of exciting wavelength (Fig. 2.13). Every group of photons absorbed

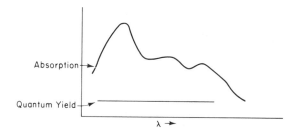

Fig. 2.13

will have the same probability of emission. These concepts are borne out experimentally. A plot of q versus wavelength is constant.

Major exceptions to these observations are azulene and rhodamine B, in which the longest wavelength absorption band do not have strong interaction with those of other wavelengths.

The constancy of quantum yield versus exciting wavelength has usefulness in elucidating the number of components in a fluorescent solution as well as use in the calibration of instruments (see Chapter 5).

2.4 Fluorescence Quenching

The fluorescence of a substance is generally more affected by the environment than is the absorption. For example, the fluorescence can be strongly "quenched" (the quantum yield lowered by substances which have little influence on the absorption spectrum). Quenching processes fall into two general types: (a) collisional quenching and (b) quenching by formation of a complex which has zero or small quantum yield. In addition, there are various intermediate states between (a) and (b). Quenchers of the first type effectively shorten the lifetime of the excited state, and since more collisions are possible the longer the excited state exists, molecules with long excited lifetimes are more sensitive to this type of quenching. Compounds forming dark complexes (b) do not affect the lifetime, but simply reduce the concentration of potentially fluorescent molecules (Weber, 1948). The formulas giving the relation between quencher concentration and fluorescence are given below:

Quenching Type (a)

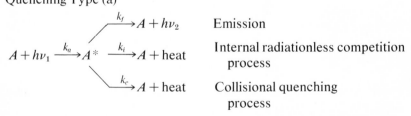

k_a, k_f, k_i, and k_c are the rate constants for the respective processes.

The number of excited molecules (A^*) are identical to the number of photons of light of energy $h\nu_1$ absorbed. These excited molecules are returned to the ground state through various mechanisms, described in the earlier section. The number of quanta emitted (fluorescence intensity) is $k_f A^*$. The total deactivation is equal to $k_f A^* + k_i A^* + k_c A^*$.

Thus the quantum yield (q) is equal to

$$\frac{k_f[A^*]}{k_f[A^*]+k_i[A^*]+k_c[A^*]+\cdots} \tag{2.9}$$

If there are no nonradiative processes then

$$q = \frac{k_f[A^*]}{k_f[A^*]} = 1 \tag{2.10}$$

In the absence of collisional quenching, the quantum yield becomes

$$q = \frac{k_f[A^*]}{k_f[A^*]+k_i[A^*]} = \frac{k_f}{k_f+k_i} \tag{2.11}$$

Note that q is independent of A^*.

We can define the observed lifetime (τ) for the excited state, when internal energy conversion is present, in terms of the rate constants

$$\tau = \frac{1}{k_f+k_i} \tag{2.12}$$

Thus it follows that

$$q = \frac{\tau}{\tau_0} \tag{2.13}$$

In the presence of collisional quenchers the rate of collisional quenching $k_c[A^*]$ must be considered:

$$q = \frac{k_f}{k_f+k_i+k_c} \tag{2.14}$$

The ratio of quantum yields in the absence and presence of quenchers becomes

$$\frac{q_0}{q} = \frac{k_f+k_i+k_c}{k_f+k_i} \tag{2.15}$$

where q_0 is the unquenched quantum yield. Combining (2.15) with the definition (2.12), we obtain

$$q_0/q = 1 + k_c\tau \tag{2.16}$$

If the concentration of the absorbing molecule A is kept constant then k_c is proportional to the concentration of the quencher:

$$k_c = kc \tag{2.17}$$

Since the fluorescence intensity is proportional to q when the exciting light intensity and concentration of absorbing molecule A are kept constant (see Chapter 5), we obtain

$$F_0/F = 1 + kc\tau \qquad (2.18)$$

where F_0 is the fluorescence intensity when no quencher is present ($c = 0$) and F is the fluorescence intensity when quencher is present ($c > 0$). This important expression is called the Stern–Volmer (1919) equation.

Quenching type (b)

$$A + Q \rightleftharpoons AQ$$
$$A + h\nu_1 \rightarrow A^* \rightarrow A + h\nu_2$$
$$AQ + h\nu_1 \rightarrow AQ^* \rightarrow AQ + \text{heat}$$

If no quencher is present, q is proportional to A^*

$$q_0 = A^*/A^* = 1 \qquad (2.19)$$

$$q = \frac{A^*}{A^* + AQ^*} \qquad (2.20)$$

$$\frac{q_0}{q} = \frac{1}{A^*/(A^* + AQ^*)} = 1 + \frac{AQ^*}{A^*} \qquad (2.21)$$

if

$$\frac{[AQ]}{[A][Q]} = K_2\text{-dissociation constant}$$

and the concentration of $Q = c$, then

$$\frac{AQ}{A} = \frac{AQ^*}{A^*} \quad \text{and thus} \quad \frac{AQ^*}{A^*} = K_2 c \qquad (2.22)$$

$$\frac{q_0}{q} = 1 + K_2 c \qquad (2.23)$$

As in Eq. (2.16) F_0 and F may be substituted for q_0 and q, respectively,

$$F_0/F = 1 + K_2 c \qquad (2.24)$$

Both types of quenching will give a straight line upon plotting F_0/F versus quencher concentration; however, types (a) and (b) can be

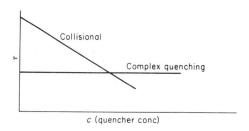

Fig. 2.14

distinguished by measuring the polarization of the fluorescence (see Fig. 2.14).

The above equations are derived from different physical models. Type (a) quenching involved consideration of the rate of the fluorescence, quenching, and internal quenching processes. Type (b) quenching was examined as an equilibrium process in which the concentration of a fluorescence molecule was altered as a function of the equilibrium constant. Since the equilibrium constant itself is formulated from two rate constants the consideration of type (b) quenching may also follow from the rate constant approach. This will not be done here. The point to be emphasized is that the processes are very similar consisting of a molecule which can be excited and a quencher fluorescent molecule complex formation which extinguishes the excited fluorescent molecule.

Equations (2.9) and (2.24) are different since the Stern–Volmer equation contains the term τ while the one for complex quenching does not. This means that if a plot were made of the concentration of quencher versus the lifetime of the excited state, then for the two processes the plots would be as in Fig. 2.14.

Another way of considering the difference in the two types of quenching is to compare them both as complexes. In the type (a) complex, the combination of fluor and quencher exists for a much shorter time than the lifetime of the excited state, whereas in the type (b) association, the two molecules are together for a period much longer than the lifetime of the excited state of the fluor alone. It is possible to have the same quencher actually yield two different types of quenching. The factors which would change for this to occur would be the lifetime of the excited state of the fluor and/or the lifetime of the fluor-quencher complex. The cases presented above are not the only ones which can occur. An intermediate type is possible, and in this phenomena the lifetime of the excited state would be similar to that of the complex. A summary of these quenching cases is presented in Table 2.1.

TABLE 2.1 COMPARISON OF QUENCHING TYPES BY
LIFETIME

Type	τ_{fluor}	τ_{complex}	$\tau_{\text{fluor}}/\tau_{\text{complex}}$
(a)	10	1	>10
	or 1	0.1	
(b)	10	100	$<1/10$
	or 1	10	
(a) + (b)	10	10	~ 1
	or 1	1	

The lifetime is expressed in arbitrary units, but for most fluors this would be between 1×10^{-9} and 1×10^{-8} sec.

2.5 Fluorescence as a Function of Concentration

The fluorescence of many solutions at high concentration of fluor is altered by a loss of quantum yield often accompanied by spectral changes. Several mechanisms are known to account for these alterations: (1) trivial reabsorption of emitted light, (2) formation of ground state dimers, and (3) formation of excited state dimers.

In the first mechanism the emitted light at the shorter wavelengths may be reabsorbed by molecules in the ground state and thus excite other molecules. This will occur if there is significant overlap of the absorption and emission spectra.

The reaction scheme for trivial reabsorption is as follows:

$$
\begin{array}{lll}
\text{i} & h\nu_1 + A_1 \rightarrow A^* & \left.\begin{array}{l} \\ \\ \end{array}\right\} \\
\text{ii} & A^* \rightarrow A + h\nu_2 & \text{Primary excitation emission} \Big\} \text{efficiency } q \\
\text{iii} & h\nu_2 + A \rightarrow A^* & \\
& \text{higher energy part} & \text{Trivial reabsorption} \Big\} \text{efficiency } q \\
& \text{of emission spectrum} & \\
\text{iv} & A^* \rightarrow A + h\nu_2 & \text{Emission}
\end{array} \qquad (2.25)
$$

If the higher energy part of the emission spectrum is absorbed once the emission efficiency would be q^2, if it is absorbed many times the efficiency of emission would be q^n where n is the number of times it is reabsorbed. Since q is less than 1 the number of quanta observed from this part of the emission spectrum would approach zero.

In the second mechanism a ground state dimer is formed between two fluor molecules

$$A + A \rightleftharpoons A_2 \qquad (2.26)$$

The concentration of monomer is obviously decreased by the amount of dimer formed

$$C_A \text{ (initial)} - 2C_{A_2} = C_A \text{ (true concentration)} \qquad (2.27)$$

If the dimer is nonfluorescent a portion of the light will be absorbed by the dimer and the fluorescence will appear to be quenched.

In many cases of ground state dimers, the absorption spectrum will change due to the dimer, in many instances shifting to longer wavelengths.

The recently discovered excited state dimer or excimer is another concentration-dependent mechanism by which some of the excited molecules are altered resulting in a decrease in the quantum yield of the monomer population (Förster, 1962).

The reaction system contains the following processes:

i	$A + h\nu_1 \rightarrow A^*$	Excitation of monomer
ii	$A^* \rightarrow A + h\nu_2$	Fluorescence of monomer
iii	$A^* + A \rightarrow D^*$	Excimer formation
iv	$D^* \rightarrow A^* + A$	Dissociation of excimer
v	$D^* \rightarrow A + A$	Internal quenching
vi	$D^* \rightarrow A + A + h\nu_3$	Fluorescence of dimer

$$(2.28)$$

The reaction scheme for excimer emission would be reactions i, iii, and vi.

After the monomer is excited there are several pathways through which the energy may be channeled depending upon the concentration of the monomer, the equilibrium constant for association of the excimer complex, and whether or not the dimer internally quenches the excitation energy or emits it at longer wavelengths.

At very low concentrations of monomer or if the association constant of the two molecules is weak then all the fluorescence observed will be that described by reaction (2.28ii).

When the concentration is increased reaction (2.28iii) is important, and a certain portion of the excited state monomer molecules will exist as the dimer, thus lowering the effective concentration of this species. The amount of this form will be a function of the rate of dissociation of the dimer (2.28iv). There are then two pathways for the excimer to lose its energy; the first by internal quenching (2.28v) and the second by emission of the fluoresence at different wavelength (normally longer). In the first case the result would be a loss of quantum yield with no other fluorescence observed. In the second, a new fluorescent band would occur. The excimer explains how spectral changes are seen at high concentrations of fluor without any changes in the ground state. Indeed the fundamental proof that the excimer is not a simple dimer is that there are no spectral changes in the ground state as the concentration of monomer is increased.

2.6 Fluorescent and Nonfluorescent Molecules

As we described earlier there are several intramolecular processes which compete for the excitation energy: fluorescence emission, internal conversion, and intersystem crossing. Intermolecular processes, i.e., collisional quenching with solvent or another solute, will deactivate the excited molecule. If the molecule dissociates in the excited state again the fluorescence may be lost.

Although all the intramolecular processes which deactivate a molecule without fluorescence emission are not known, certain examples of substituted systems have yielded data consistent with quantum mechanical models. We have, in the previous chapter, considered the excited state of a molecule as only singlet or triplet. In a strict sense this is not true. The singlet state has some triplet character. Thus there can be intersystem crossing so that the excited singlet state can become a triplet one. Since the triplet-ground state conversion is forbidden, other deactivation processes such as solvent quenching may have a longer time to effect the molecule.

Halogen substituted naphthalene derivatives show decreased fluorescence with increased phosphorescence. These derivatives have a corresponding increased triplet character in the excited state. This same concept of increased intersystem crossing or electron reversal is the reason given for the loss of fluoresence following and as a substitution. Similarly when heavy atoms are introduced into fluorescent dyes, they are often quenched.

The effect of the solvent on quenching processes is very important and is described in detail in the next chapter. In part, the solvent acts by converting the excited molecule to the ground state either through direct loss of energy to the solvent or by absorbing the energy through a series of internal vibrational steps with the solvent as the final recipient.

2.7 Chemiluminescence and Bioluminescence

So far we have discussed excited states induced by the absorption of photons. Excitation may, however, be of a purely chemical origin, and the associated luminescent phenomenon are then termed bio-luminescence or chemiluminescence depending on whether they occur in biological systems or not. It is not our intention to give a general presentation of these phenomena or to go into details of the various chemical mechanisms leading to luminescence; the interested reader will find such accounts listed among the references at the end of this chapter. Rather, we will give some examples of chemical excitation and briefly discuss the energetics behind these types of luminescence.

Some of the reactions which are luminescent are presented in Table 2.2. These are somewhat arbitrarily divided into chemical, enzymatic

TABLE 2.2 SUMMARY OF CHEMILUMINESCENT AND BIOLUMINESCENT REACTIONS[a]

Chemical

$$R \cdot + R \cdot \longrightarrow R - R + h\nu \text{ (single bond formation)}$$
$$\cdot R \cdot + \cdot R \cdot{'} \longrightarrow R = R + h\nu \text{ (double bond formation)}$$
$$RO_2 \xrightarrow{\Delta} \cdot R \cdot + O_2 \longrightarrow R \cdot + h\nu \text{ (electric spark discharge)}$$
$$R^+ + e^- \longrightarrow R + h\nu \text{ (electron capture)}$$

Enzymatic-nonspecific

$$\text{Substrate or } L + O_2 \xrightarrow{\text{peroxidase}} \text{oxidized } L + h\nu$$

Bioluminescent

$$FMN - H_2 + RCHO + O_2 \xrightarrow{\text{luciferase}} FMN + h\nu + \text{products}$$
$$LH_2 + ATP + O_2 \xrightarrow{\text{luciferase}} L + AMP + H_2O + h\nu$$
$$LH_2 + O_2 \xrightarrow{\text{luciferase}} L + H_2O$$
$$LH_2 + H_2O_2 \xrightarrow{\text{peroxidase}} L + H_2O + h\nu$$

[a]This table is compiled from Paris (1966), Degn (1969), and Cormier and Totter (1964).

(nonspecific), and bioluminescent (specific reactions occurring in living organisms). Four generalized chemical reactions which result in the production of light are presented in the first part of the table (Paris, 1966). The requirements for a chemiluminescence reaction of this type are that the excited state of the product molecule has a high enough energy and that there is a transition from this excited state to the ground state by the emission of a photon which is competitive with other mechanisms of deactivation, i.e., the excited molecule should be fluorescent.

Chemiluminescence has also been found in several reactions cata-lyzed by peroxidase in the presence of O_2, but not H_2O_2 with substrates such as indole-3-acetic acid. The quantum efficiency of this reaction has not been measured, but the chemiluminescence has been ascribed to the reactions of the free radical form of indole-3-acetic acid. However, for a similar reaction the chemiluminescence has been said to be a product of decomposition of one of the enzyme intermediates (Degn, 1969). In another study using a system with a higher quantum efficiency (i.e., luminol, hydrogen peroxide, and peroxidase); no luminescence was found from decomposition of enzyme intermediates, rather it was from the reaction of two free radicals with hydrogen peroxide (Cormier and Prichard, 1968).

$$E + H_2O_2 \longrightarrow C_1 \qquad\qquad (2.29)$$

$$C_1 + LH_2 \longrightarrow C_2 + LH\cdot \qquad\qquad (2.30)$$

$$C_2 + LH_2 \longrightarrow E + LH\cdot \qquad\qquad (2.31)$$

$$2LH\cdot + H_2O_2 \longrightarrow h\nu + \text{products} \qquad\qquad (2.32)$$

where L is luminol

These reactions are of interest since they can occur at room temperature in aqueous solutions. The light production in this case requires that the free radicals generated be stable long enough for two radicals to react. A luminescent reaction also takes place with peroxidase, H_2O_2, and pyrogallol. Sensitive methods for peroxidase assay have been found in that reaction (Ahnström et al., 1961; Ahnström and Nilsson, 1965).

Bioluminescence has been found in a large number of organisms (Table 2.3), and the reactants responsible for the luminescent reactions have been described in detail in several cases. The most carefully studied reaction sequence is that of the firefly (Fig. 2.15), which involves an enzyme (luciferase), a coenzyme (luciferin, activated by ATP), and a substrate (molecular oxygen). The structure of luciferin presented here is for the firefly. Other luciferins which differ structurally have been found. However, it is interesting that the same type of molecule can occur in organisms as diverse as the firefly and dinoflagellates. In contrast to the other systems presented, there is no evidence for a free radical mechanism. The system is, however, unusual in that it involves reaction of molecular oxygen with no hydrogen peroxide intermediate. Thus all the energy of this oxidation reaction

TABLE 2.3 TYPES OF BIOLUMINESCENT SYSTEMS[a]

Organisms	Reaction system	Reactions causing luminescence	Max. of the emission (mμ)
Bacteria (*Photobacterium fischeri* and others)	Pyridine nucleotide	1) $NAD\text{-}H_2 + FMN \underset{}{\overset{NAD\text{-}H_2\text{-oxidoreductase}}{\rightleftharpoons}} FMN\text{-}H_2 + NAD$ 2) $FMN\text{-}H_2 + RCHO + O_2 \xrightarrow{E}$ light	465–495
Fungi (*Omphali flavida* and others)	Pyridine nucleotide	1) $NAD\text{-}H_2 + L \underset{}{\overset{NAD\text{-}H_2\text{-oxidoreductase}}{\rightleftharpoons}} LH_2 + NAD$ 2) $LH_2 + O_2 \xrightarrow{E}$ light	530
Insects (*Photinus pyralis* and others)	Adenine nucleotide	$LH_2 + ATP + O_2 \xrightarrow[Mg^{2+}]{E}$ light	562
Corals (*Renilla reniformis*)	Adenine nucleotide	$LH_2 + DPA + O_2 \xrightarrow{E}$ light	Blue region
Crustacea (*Cypridina hilgendorfii* and others)	Enzyme substrate		460
Dinoflagellate (*Gonyaulax polyhedra*)	Enzyme substrate		470
Worms (*Odontosyllis enopla* and others)	Enzyme substrate	$LH_2 + O_2 \xrightarrow{E}$ light	510
Mollusks (*Pholas dactylus*)	Enzyme substrate		480
Fishes (*Parapriacanthus beryciformis* and others)	Enzyme substrate		460
Enteropneusta (*Balanoglossus biminiensis*)	Peroxidase	$LH_2 + H_2O_2 \xrightarrow{E}$ light	Blue region
Jellyfish (*Aequorea aequorea* and others)		Protein $\xrightarrow{Ca^{2+}}$ light	460 and 510

[a]LH_2 stands for reduced luciferin, E for Luciferase, DPA for 3′, 5′ diphosphoadenosine. The table is taken from Barenboim et al. (1969) and compiled mainly from McElroy and Seliger (1962b, c) as well as from Cormier and Totter (1964) and Chase (1964).

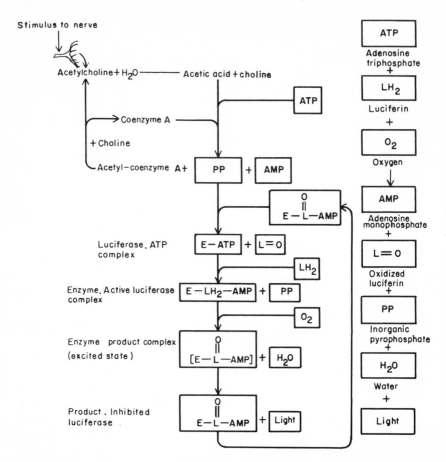

Fig. 2.15. Firefly flash is probably triggered by a nerve impulse delivered to the luminous gland. A sequence of chemical reactions then produces light. The substances consumed in the reaction, as shown in the summary at right, are adenosine triphosphate (ATP), luciferin (LH_2), and oxygen (O_2). The products are oxidized luciferin (L=O), two phosphate compounds, water and light. The reaction is catalyzed by the enzyme luciferase, represented by E. One quantum of light is produced for each molecule of luciferin oxidized. From McElroy and Seliger (1962a).

appears to excite the luminescent molecule which then has a very high probability (approaching unity) to revert to the ground state by the emission of a photon.

A more detailed sequence of the reaction of the coenzyme is presented in Fig. 2.16. It appears that the coenzyme ionizes before emitting a photon. As we shall see in the next chapter, the ionization of a molecule in the excited state is a common phenomenon.

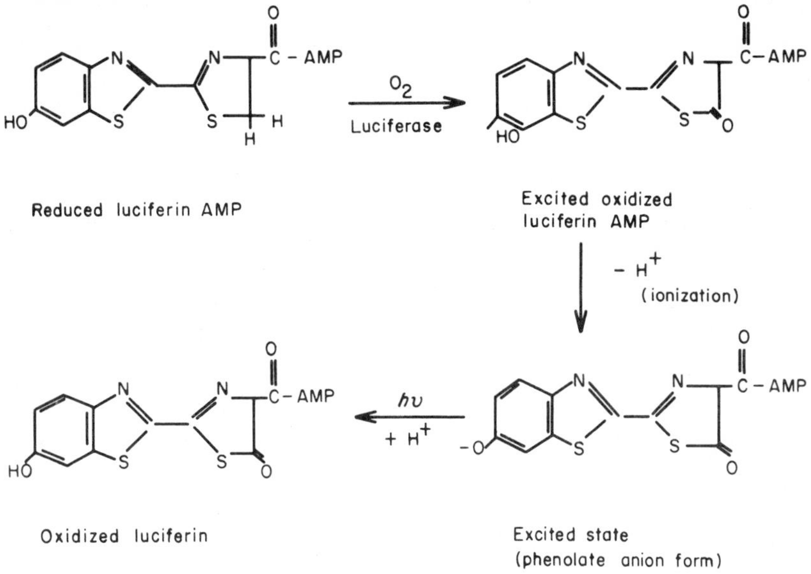

Fig. 2.16. Proposed detailed reaction sequence for luciferin bound to luciferase resulting in oxidation and production of a photon.

The reaction thus requires luciferase, luciferin, ATP, O_2, and also Mg^{2+}. With the concentrations of all other components fixed, the variation of any one of them may be recorded as a variation in luminescence intensity. Since the ATP content is a fairly constant characteristic of any species of microorganism, the firefly luminescent system may also be utilized for the determination of microbial content of a fluid. There is commercial equipment available, which gives a direct readout of either ATP content or number of cells, but regular fluorometric equipment may easily be converted for such use.

One common characteristic of chemiluminescence or bioluminescence reactions is the requirement for molecular oxygen or hydrogen peroxide. Another one is the existence of a ring system involving nitrogen and/or keto groups, which have mobile electrons taking part in the $n \to \pi$ transition, which constitutes the excitation.

The emitted light is usually in the blue to green range, i.e., the maxima are in the 470–570 nm in wavelength interval. This corresponds to energies of about 50–60 kcal/mole. Since this energy is far in excess of the energy which is released in the splitting of a so-called energy-rich bond, e.g., $ATP \to AMP + PP$ (8 kcal/mole), it is realized that the excitation energy involved in a bioluminescent process cannot be

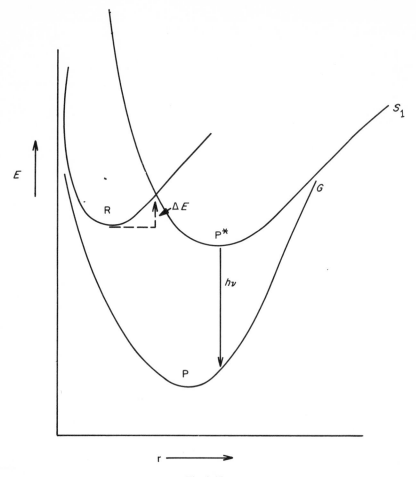

Fig. 2.17.

supplied in one single step. Rather, it must be built up through a complex series of chemical steps as we have seen in the examples above.
That chemical reaction which immediately precedes the light emission
may simply be a triggering step as indicated in the energy level diagrams of Fig. 2.17 pertaining to the reaction sequence $R + \Delta E \rightarrow$
$P^* \rightarrow P + h\nu$, where R is reactant, ΔE is the energy barrier, P^* the
excited product, and P the ground state product.

REFERENCES

Ahnström, G., and Nilsson, R., *Acta Chem. Scand.*, **19**, 313 (1965).
Ahnström, G., v. Ehrenstein, G., and Nilsson, R., *Acta Chem. Scand.*, **15**, 1417 (1961).

Barenboim, G. M., Domanskii, A. N., and Turoverov, K. K., *Luminescence of Bio-polymers and Cells*, Plenum Press, New York, 1969, p. 126.

Berlman, I. A., *Handbook of Fluorescence Spectra of Aromatic Molecules*, Academic Press, New York, 1965.

Birks, J. B., and Dyson, D. J., *Proc. Royal Soc. (London)* **A275**, 135 (1963).

Chase, A. M., in *Photophysiology* (A. C. Geise, ed.), Vol. 2, Academic Press, New York, 1964, p. 384.

Condon, E. V., and Morse, P. M., *Quantum Mechanics*, McGraw-Hill, New York, 1929, pp. 164ff.

Cormier, M. J., and Prichard, P. M., *J. Biol. Chem.*, **243**, 4706 (1968).

Cormier, M. J., and Totter, M. J., *Ann. Rev. Biochem.*, **33**, 431 (1964).

Degn, H., *Biochem. Biophys. Acta*, **180**, 271 (1969).

Förster, Th., *Fluoreszeng Organischer Verbindungen*, Vandenhoeck Ruprecht, Göttingen, 1951, p. 154.

Förster, Th., *Molecular Spectroscopy* (5th European Congress on Molecular Spectroscopy), Butterworth, London, 1962, p. 121.

Hercules, D. M., ed., *Fluorescence and Phosphorescence Analysis*, Wiley (Interscience), New York 1966, pp. 25ff.

Levshin, W. L., *Z. Physik*, **43**, 230 (1931).

McElroy, W. D. and Seliger, H. H., *Sci. Am.*, **207**(6), 76 (1962a).

McElroy, W. D., and Seliger, H. H., in *Horizons in Biochemistry*, Academic Press, New York, **91**, (1962b).

McElroy, W. D., and Seliger, H. H., in Proceedings of the Fifth International Congress, Symposium *3*, *Evolutionary Chemistry*, Moscow, 1962c, p. 161.

Morton, R. A., Hopkins, T. A., and Seliger, H. H., *Biochem.*, **8**, 1598 (1969).

Paris, J. P., in *Fluorescence and Phosphorescence Analysis*, (D. M. Hercules, ed.) Wiley (Interscience), New York, 1966, p. 185.

Seybold, P., and Gouterman, M., *Chem. Rev.*, **65**, 413 (1965).

Stern, O., and Volmer, M., *Z. Physik.*, **20**, 183 (1919).

Weber, G., *Trans. Faraday Soc.*, **44**, 185 (1948).

Weber, G., and Young, L. B., *J. Biol. Chem.*, **240**, 1415 (1964).

BIBLIOGRAPHY

Barrow, G. M., *Introduction to Molecular Spectroscopy*, McGraw-Hill, New York 1962.

Guilbault, G. G., *Fluorescence—Theory, Instrumentation and Practice*, Marcel Dekker, New York, 1967.

Hercules, D. M., ed., *Fluorescence and Phosphorescence Analysis*, Wiley (Interscience), New York, 1966.

Johnson, F. H., in *Bioluminescence* and Florkin M., and Stotz, E. H., *Photobiology, Ionizing Radiation*, Vol. 27 of Comprehensive Biochemistry, Elsevier, Amsterdam, 1967.

Udenfreund. S., *Fluorescence Assay in Biology and Medicine*, Academic Press, New York, 1962; and Vol. II (1969).

Chapter 3

Solution Effects on Absorption and Emission Spectra

3.1 Introduction

A knowledge of possible environmental effects on spectra and quantum yields of fluorescence is necessary for the utilization of fluorescence technique at its maximum potential. Since such effects are most important in liquid systems, and since most biologically oriented workers are primarily interested in the fluorescence of solutions, we will give these factors special consideration. A list of the environmental factors that affect fluorescence phenomena includes interactions with solvent and other dissolved compounds, temperature, pH, and the concentration of fluorescent species. The effects that these four parameters have upon fluorescence vary from fluorescent moiety to moiety. Both the absorption and emission spectra as well as the quantum efficiencies of fluorescent molecules are influenced by these parameters. Although this influence may often be a complication to the fluorescence spectroscopist, a variation of the above-mentioned factors may often give important information concerning the nature of a fluorescent compound. In this way, it is possible to study the effects that different constituent groups of a family of fluorescent substances have upon the fluorescence of the compound as a whole, e.g., the family including anisole and tyrosine of which phenol is the parent compound. Conversely, variation of environmental factors may in many cases make it possible to distinguish between different fluorescent moieties with similar spectra. When considering the effects of varying solvents upon the fluorescence of a compound it is impossible to isolate the influence upon emission processes from the effects upon the absorption processes. This is due to limitations in current methods which cannot define changes in the fluorescence spectra occurring in the excited state without also considering the ground state of the molecule. (Normally consideration of the ground state changes are done by evaluating differences in the absorption spectra of the molecule.) Any spectral changes that occur due to changes in solvent, temperature, pH, and concentration are results of phenomena occurring either in the ground state or in the excited state or both. Those changes could be the result of many interactions, i.e., complex formation which would alter the energy levels, quenching of the fluorescence, or dipole–dipole interactions between the solvent and the fluorescent compound in the ground or the excited state. All of these phenomena are important in defining the causes for shifts in the fluorescence spectra, the absorption spectra, or for changes in the quantum yield of fluorescence.

The large range of concentration of solute which may be studied by fluorescent techniques (high sensitivity) is due to the direct measurement of light intensity as opposed to differences in light intensity as utilized in absorption spectroscopy (see Chapter 6 for relationship between intensity and fluorescence or absorption). Therefore, fluorescence is a very good technique for the study of environmental effects upon both the excited state and the ground state of a molecule over a large concentration range. Investigation of changes in the ground state of solute molecules is in this case accomplished by use of the excitation spectrum of fluorescence at varying concentrations of solute. In this way it is possible to examine the absorption characteristics of a fluorescent molecule when in the presence of other absorbing, nonfluorescent solutes. This is often impossible by absorption spectrophotometry.

3.2 Solvent Effects

A. Spectral Shifts

It has long been observed that many compounds will have shifts in their uv absorption spectrum when the solvent is varied. Similarly the fluorescence spectrum of many compounds will change as the solvent is varied. The variation in solvent may be: (1) a change from polar to nonpolar; (2) the reverse; (3) a change in the dielectric constant; (4) a change in the polarizability (Barrow, 1966). In many cases not only a shift in the fluorescence spectrum is observed but this is often accompanied by a shift in the absorption spectrum of the compound. However, a shift in one does not necessarily require a shift in the other. It is common for a shift in fluorescence to occur with no shift in the absorption spectrum.

Since the excited state is different from the ground state (see Chapter 1), i.e., it may have a greater dipole moment, then the ground state of a particular molecule may change very little from solvent to solvent, but the excited state energy may be quite different because of its dipole moment. Thus there may be little spectral change in the ground state of the molecule but a large one in the emission spectrum as the solvent is altered.

Figure 3.1 is a representation of some hypothetical energy levels for a model molecule. E_1 is the energy of transition from the zero vibrational ground level to the zero vibrational, first excited singlet level in

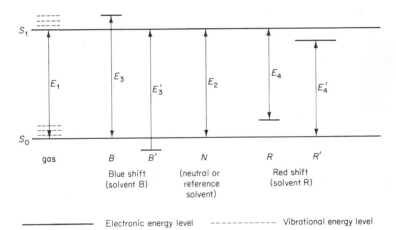

Fig. 3.1. Representation of energy levels for a hypothetical molecule in environment giving rise to spectral shifts. As transitions may occur in either direction ($S_1 \rightleftharpoons S_0$), shifts in both emission and absorption spectra are obtained.

the gaseous state (referred to as a 0–0 transition). As depicted, this would represent an unperturbed molecule. Unperturbed is defined as the state of the molecule which occurs at very low pressure in the gas state, i.e., there are no interactions between molecules. The absorption spectrum here depends entirely upon the internal energy levels of the molecule. Thus the absorption spectrum is a function of only one external parameter, the temperature, which affects the distribution (Boltzmann distribution) of molecules in the various vibrational states in the ground singlet level. However, this vibrational distribution is very small at normal temperatures for organic molecules. For comparison there is a diagram for the same molecule in a neutral solvent, in which we assumed that the singlet–singlet transition energy E_2 is the same as E_1 of the gas. The absorption spectrum will not be the same in solution as in the gas phase since interactions with the solvent will cause band broadening.

First let us consider the case where the solvent changes to one which causes the mean of the absorption band to move toward longer wavelengths (a red shift), which means a lower energy difference between the ground and excited states. This change can be accomplished in one of two ways: (1) an interaction occurring in the ground state which would raise the ground state energy level and would not affect the excited state level as represented by R in Fig. 3.1; (2) an interaction occurring in the excited state, but not in the ground state as represented in Fig. 3.1 by R'. A shift in the mean of the absorption spectrum to shorter wavelength or higher energy (blue shift) can be accomplished by placing the molecule in a solvent which will alter the difference in the transition energy required to excite the molecule. This is illustrated in case B of Fig. 3.1. The excited state undergoes some interaction with the solvent in which its energy is raised, the ground state remaining at the same energy level. Therefore, the required energy for absorption of E_3 is greater than that of E_2, the unperturbed state, resulting in a blue shift. Another way this could occur is by the mechanism of case B' of Fig. 3.1 in which the excited state undergoes no interaction with the solvent but the ground state is altered so that its energy is below that of the ground state in the neutral solvent. In this case the energy of absorption E_3' is greater than E_2, yielding a blue shift. It is not always possible to determine unambiguously which of the two types of phenomenon, case B or B' of Fig. 3.1, is responsible for blue or red spectral shifts. In fact, in many molecules some of both phenomena actually occur. To understand more fully this physics of light excitation of a molecule an understanding of the Franck–Condon principle is essential. This principle states (Chapter 1) that upon excitation of a molecule the electron which is raised to a new electronic level is excited in much less time than it takes the whole molecule to rearrange itself with the solvent environment. Immediately upon absorption of energy causing the electronic transition, the molecule is in the same structural environment in the excited state as in the ground state. For this reason the excited state interaction with the solvent occurring immediately upon photon absorption is not the one seen at a later time. In other words, if there is a certain arrangement of the ground state molecule with the molecules of solvent, i.e., mutual orientations of solvent and solute molecules, immediately upon excitation this arrangement has not changed. It is, however, not necessarily the most stable arrangement for this complex of excited state molecules and solvent molecules. After a certain time interval has passed, a most probable excited state interaction with the solvent environment will evolve and this may not necessarily have the same energy level as that existing at the time the molecule was first excited. To help clarify this situation see Fig. 3.2,

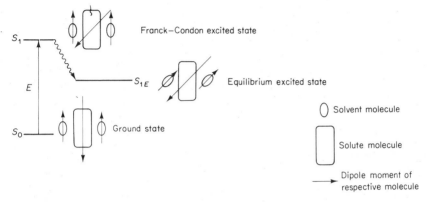

Fig. 3.2. Franck–Condon effect. Demonstrated for dipole–dipole interaction between solvent and solute. S_0 = ground state. S_1 = Franck–Condon excited state. S_{1E} = equilibrium excited state. The ordinate is of increasing energy.

where we have depicted an energy level diagram for a molecule in which the initial absorption occurs from the ground state equilibrium energy level to the "Franck–Condon excited state." After a certain time period a solvent–solute equilibrium is established so that the Franck–Condon excited state no longer remains, but now there is an excited state in which the molecule is in an equilibrium energy level resulting from a preferred state of orientation and polarization vis-á-vis the solvent. Figure 3.2 presents a simple example of dipole–dipole interactions in which the solute molecules have a given dipole moment with respect to the molecular axis. The dipole moment is assumed to be different in the ground and excited states. The illustration shows the solvent molecules as dipoles with given orientations with respect to the dipole of the solute. Upon excitation, which occurs in approximately 10^{-15} sec, the excited molecule has a dipole moment which is oriented differently from that of the ground state. Within the 10^{-15} sec the solvent dipole molecule has not had time to reorient itself to the lowest energy orientation with respect to the excited state dipole. After approximately 10^{-10} sec the solute molecules have had a chance to rearrange themselves into the most favorable arrangement of alignment with the excited state dipole. This results in a lowering of the excited state energy level.

Both examples of Fig. 3.3 are further illustrations of the Franck–Condon principle, in which the solute molecules have no (or very small) dipole moments in the ground state, but do possess a dipole moment in the excited state. For purposes of simplicity we can neglect any secondary interactions between solvent and solute such as dipole–dipole interaction in the ground state, or any interactions such as hydrogen bonding and van der Waals attraction.

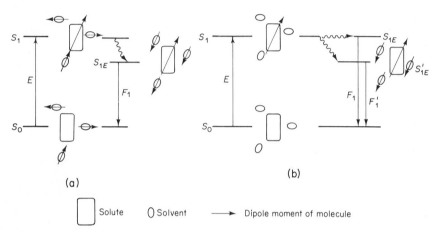

Fig. 3.3. Franck–Condon effect in various solvents. The solute molecule in this instance has no dipole moment in the ground state ($\mu_g = 0$). E = excitation, F_1 = fluorescence. (a) The solute is in a polar solvent. S_0, S_1, and S_{1E} as in Fig. 3.2, the ordinate being of increasing energy. (b) The solute is in a nonpolar solvent. Two cases arise here depending upon the polarizability of the nonpolar solvent and also the magnitude of μ^* of the excited solute molecule. When S_{1E} obtains above, either the solvent is of too low polarizability, or μ^* is too weak, or both such that there is no induced-dipole moment in the solvent. When S'_{1E} obtains, the magnitudes of μ^* and solvent polarizability are such that a dipole is induced in the solvent allowing dipole-induced dipole interaction and change in the equilibrium excited state.

In Fig. 3.3a the solute is in a polar solvent. Therefore the "Franck–Condon excited state" is higher than the equilibrium excited state since once the molecule attains the excited state it becomes a dipole. This necessitates some rearrangement before equilibrium between the solvent dipole and the solute dipole is reached. In Fig. 3.3b the solvent is nonpolar and the ground state energy is assumed to be the same as in Fig. 3.3a. Here the Franck–Condon excited state is lower than in the polar solvent because of the lack of dipole–dipole interactions. Since there is no dipole–dipole incorrect orientation in the excited state, the initial excited state will be of approximately the same energy as the equilibrium excited state. An intermediate case would exist if the non-polar solvent were of sufficient polarizability so that the excited state dipole can induce dipole moments in the solvent molecules. In this instance, after equilibrium has been reached between the excited solute dipoles and the induced dipoles of the solvent, the excited state energy level will be lower than that of the Franck–Condon excited state. Figure 3.3 also explains the relationship of spectral shifts to solvent–solute interaction phenomena. The blue shift in the absorption spectrum of a solute as the solvent is changed from the nonpolar to polar results from the initially unfavorable interaction between the excited state dipole and the solvent dipoles of the polar solvent. The direction of emission spectral shifts are in contrast and are less easy to predict from the polarity of the solvent. If the difference between the Franck–Condon excited state and the equilibrium excited state for the polar solvent $[S_1 \text{ (polar)} - S_1' \text{ (polar)}]$ is greater than the difference between the Franck–Condon excited state in the polar solvent and the equilibrium excited state in the nonpolar solvent $[S_1 \text{ (polar)} - S_1' \text{ (nonpolar)}]$, the fluorescence as one goes from polar to nonpolar will undergo a blue shift (Fig. 3.4a). If the reverse is true, the shift in the fluorescence spectrum will be to the red (Fig. 3.4b), and the energy of the

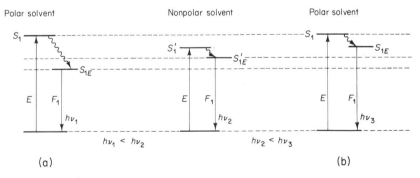

Fig. 3.4. Fluorescence solvent shifts. Models showing shifts as a function of the energy difference between Franck–Condon and equilibrium excited states. The assumption has been made that $S_0 \longrightarrow S_1$ in polar solvent is the same for both cases and that $S_0 \longrightarrow S_1$ in nonpolar solvent is less than in the polar solvent. $E = $ excitation, $F_1 = $ fluorescence. S_0 is ground state. S_1 and S_{1E} are Franck–Condon and equilibrium excited states in polar solvent. S_1' and S_{1E}' are Franck–Condon and equilibrium excited states in nonpolar solvent. $h\nu_1$ and $h\nu_3$ are the energies of $S_{1E} \longrightarrow S_0$ transitions in the two polar solvents. $h\nu_2$ is the energy of $S_{1E}' \longrightarrow S_0$ transition in the nonpolar solvent. (a) Blue shift in emission going from polar to nonpolar solvent. $E_{S1} - E_{S1E} > E_{S'1} - E_{S'1E}$ where E_S is the energy of state S. $h\nu_1 < h\nu_2 \equiv \lambda_1 > \lambda_2$; thus a blue shift ($\lambda = $ wavelength of emission $-\lambda = c/\nu$). (b) Red shift in emission going from polar to nonpolar solvent. $E_{S'1} - E_{S1E} < E_{S'1} - E_{S1E}'$. $h\nu_3 h\nu_2 \equiv \lambda_3 < \lambda_2$; thus a red shift.

equilibrium excited state in the nonpolar solvent is less than the energy of the equilibrium excited state in the polar solvent. For fluorescence emission spectra only, the equilibrium and not the Franck–Condon excited state is considered, thus allowing for two different emission spectra in the case of Fig. 3.3b where two equilibrium excited states are obtained depending upon whether the nonpolar solvent is polarizable and the magnitude of its polarizability. From the simple model depicted in Fig. 3.4 it is not possible to predict precisely whether there will be a blue shift or red shift in the fluorescent spectrum of the solute molecule.

A scheme more appropriate for description of actual phenomena is shown in Fig. 3.5. In this case, consider that there is a phenomenon in emission similar to the Franck–Condon principle in absorption. After

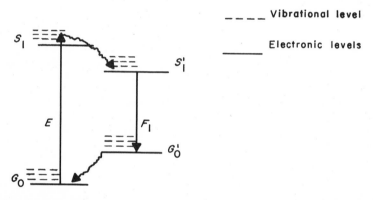

Fig. 3.5. A more realistic picture of excited states of a molecule. G_0 = ground state. S_1 = Franck–Condon excited state, S_1' = equilibrium excited state. G_0' = Franck–Condon equivalent ground state after emission. $S_1 \longrightarrow S_1'$ and $G_0' \longrightarrow G_0$ are radiationless transitions (internal conversion or thermal relaxation) to lower energy states dictated by solvent–solute interactions. E = excitation, F_1 = fluorescence. Ordinate is increasing energy.

the transition has occurred, there follows some solvent–solute rearrangement to reach an energetically preferred state. Another way of pointing this out is to compare the time parameter. Even though the *lifetime of the excited state* (S_1') may be on the order of 10^{-10} to 10^{-8} sec, *the transition time for the electron to go from S_1' to G_0', once elected to undergo transition, is still on the order of 10^{-15} sec,* so that the Franck–Condon principle holds for *emission* as well as absorption. It must be emphasized that nonradiative decay processes from the excited state to the ground state do not obey the Franck–Condon principle and may or may not form any of the excited energy states of

Fig. 3.5 (nonradiative decay could occur from S_1' to G_0' directly or through undefined levels, probably not including G_0').

It may be pointed out that absorption spectral shifts to longer wavelengths (red shifts) result from induced polarization of solvents due to the transition dipole of the solute and subsequent formation of induced-solvent dipoles, that is, the orientation of the solvent dipoles is influenced by the dipole of the electronic transition when it is excited. Since it is a two-way interaction the transition dipole is influenced by the solvent. Those solvents which interact more strongly with the transition moment change the absorption spectra. This is most clearly seen in nonaqueous solvents and solutes in which the red shift occurs as the dielectric constant of the solvent increases. This is sometimes obscured by other effects, however, such as hydrogen bonding and van der Waals interactions. [See Bayless and Mehar (1954) for a further discussion on solvent effects and their origin.] Blue shifts in the absorption spectra of solutes in solution of high dielectric constants have been discussed in terms of the Franck–Condon principle (McConnel, 1952; Pimentel, 1957). These shifts have been attributed to the difference in solvation energy of the solute in the ground state and the excited state. Again, use is made of the energy changes between the ground and excited states as illustrated in Figs. 3.3 and 3.4 to get some idea of the relationships between the solvation energy, the Franck–Condon principle, and spectral shifts.

Hydrogen bonding has been stressed in solvent effects (Pimentel, 1957; Brealey and Kasha, 1955; Matoge and Tsuno, 1957) and is especially important in oxygen and nitrogen donors or acceptors. This ability varies greatly between the excited and ground states. Hydrogen donors and acceptors will be discussed later in the section on the effect of pH on fluorescence spectra particularly in the more drastic cases of ionization of constituent groups on the molecule. It is also possible for the excited state to have contributions from structures not possible in the ground state with resulting greater sensitivity to solvent changes. An example of this is indole (Fig. 3.6a) which has little or no changes in absorption spectrum with increasing dielectric constant of the solvent, but an emission spectral shift to longer wavelengths (Van Duuren, 1961). This has been explained as resulting from a greater contribution of mesomeric forms of lower energy as shown in Figs. 3.6b and 3.6c to the excited states. The formation of these structures is facilitated by an increase in the dielectric constant of the solvent. Since indole is neither a hydrogen donor nor a hydrogen acceptor, hydrogen bonding may be eliminated as a possible explanation for these shifts. Table 3.1 shows some data for solvent effects on spectra of different compounds.

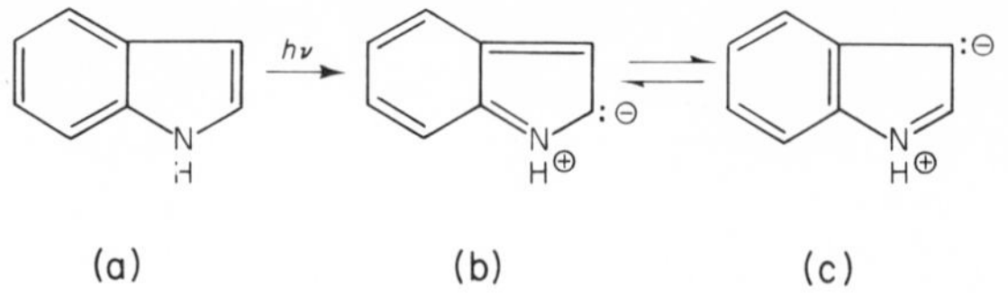

(a)   (b)   (c)

Fig. 3.6. Indole excited states. (a) Indole. (b) and (c) Mesomeric forms of indole thought to contribute to the excited state structure of the molecule and giving rise to the shifts in emission spectra in solvents of varying dielectric constant.

B. QUENCHING

Quenching of fluorescence is defined as a decrease in the quantum yield of fluorescence. (Quantum yields are discussed in Chapter 2, the rate processes describing quantum yields in quenching are covered in Sec. 2.1.) The mechanism of solvent quenching cannot be described in all cases, but some examples where quenching may occur are: (1) increase of the rate of intersystem crossing (from the excited singlet to the excited triplet) thus increasing phosphorescent yield; (2) formation of nonfluorescent complexes via hydrogen bonding, van der Waals interaction, or dipole–dipole interaction; (3) ionization to a non-fluorescent species. Van Duuren (1962) includes a review of specific solvent quenching effects with references. Table 3.2 lists some specific quantum yields of compounds in various solvents.

3.3 Effect of pH

The degree of protonation of a molecule, dependent upon its electronic structure as well as the constituent groups present in the molecule, may change upon excitation. The change in protonation with absorption is evidenced by changes in the fluorescence or absorption spectrum of the molecule with varying pH. For a change in pK of the ionizable groups to be seen, ionization equilibrium must be obtained during the lifetime of the excited state. [For a more detailed discussion of energy changes in fluorescence and absorption spectra, with examples given, see Weller (1957).] Figure 3.7 is an energy diagram for the dissociation of an acid HA and its base A (Van Duuren, 1962). ΔH is the enthalpy of dissociation of the ground state, ΔH^* is the enthalpy of dissociation of the excited state. If ΔH is greater than ΔH^* a red shift occurs in the absorption and fluorescence on acid dissociation. If ΔH is less than ΔH^* then a blue shift occurs in absorption and

TABLE 3.1 SOLVENT EFFECTS ON SPECTRA OF AROMATIC COMPOUNDS

Solvent compound	Cyclohexane			Benzene			Ethyl alcohol			Water		
	λ_{ex}[a]	λ_{EM}[b]	F[c]	λ_{ex}	λ_{EM}	F	λ_{ex}	λ_{EM}	F	λ_{ex}	λ_{EM}	F
Quinoline[d]	317	330	Very weak			0.001		305 370	0.03			
Acridine[d]	250 365	400	Very weak	—	—	—	250 340 355 380	415 440	1	250 355 382	430 455 475	8
5,6 Benzquinoline[d]	273 324 337	344 361 379	1	278 320 333	345 362 380	1	275 286 315 329 342	348 365 380	3.2	272 285 315 329 342	350 365	4.0
Indole[e]	285	297	6.0	285	305	0.6	285	330	6.1	285 280	350[f] 350	4.0
2-Methylindole[e]	280	306	6.2	280	305	Very weak	280	335	6.5	280	355[f]	2.0
1-Methyl-2-phenyl indole[e]	310	360	64.0	310	370	63.0	310	370	61.0	305	380	17.0
Tryptophan	—	—	—	—	—	—	—	—	—	280	345[g] 348[h] 360[i]	—

[a]Wavelength of maximum excitation of fluorescence (nm).

[b]Wavelength of maximum emission (nm).

[c]Relative fluorescence intensity.

[d]From Van Duuren, B. L., *Abstr. Pittsburgh Conf. Anal. Chem. Appl. Spectry.,* **11**, 61 (1960); *Anal. Chem.,* **32**, 1436 (1960).

[e]Except where elsewhere specified, data taken from Van Duuren, B. L., *J. Org. Chem.,* **26**, 2956 (1961) at room temperature.

[f]Sprinc, H., Rowley, G. R., and Jamieson, D., *Science,* **125**, 442 (1957).

[g]At $-196°$: in presence of 0.5 to 10% glucose in water this maximum is shifted to 325 nm and doubled in intensity. Fujimori, E., *Biochim. Biophys. Acta,* **40**, 251 (1960); Steele, R. H., and Sjent–Gijorgyi, A., *Proc. Natl. Acad. Sci. U.S.,* **43**, 477 (1957).

[h]Teale, F. W. J., and Weber, G., *Biochem. J.,* **65**, 476 (1957).

TABLE 3.2[a] QUANTUM YIELD OF FLUORESCENCE OF AROMATIC HYDROCARBONS AND DERIVATIVES IN VARIOUS SOLVENTS[b]

	Hexané	Acetic acid	Benzene	Toluene	p-Xylene	Pyridine	Chlorobenzene	Thiophene	Chloroform
Anthracene	0.32	0.31	0.24 0.241[c]	0.23	0.17	0.16	0.15	0.12	0.10
1-Chloroanthracene			0.09						0.05
1,5-Dichloroanthracene			0.065						0.04
9,10-Dichloroanthracene			0.65						0.50
9-Phenylanthracene			0.74						0.40
9,10-Diphenylanthracene			0.80 0.840[c]						0.65
Perylene			0.06 0.800[c]						0.88

[a] Table 3.2 from Van Duuren, Chem. Rev., 63, 325 (1962).

[b] At 20° and 365 excitation except for perylene excited at 313.5. [Bowen, E. J., Trans. Faraday Soc., 50, 97, (1954); Bowen, E. J., and Stebbins, D. M., J. Chem. Soc., 1957, 360; Bowen, E. J., and West, K., J. Chem. Soc., 1955, 4394. Quantum yields are obtained from assumption of absolute quantum yield of anthracene in benzene to be 0.24, and direct conversion of yields relative to anthracene into absolute yields.

[c] Melhuish, W. H., J. Phys. Chem., 65, 229 (1961).

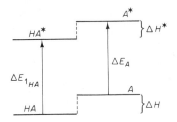

Fig. 3.7. Energy diagram for dissociation of acid (HA) in the ground and excited states (Van Duuren, 1963).

fluorescence upon acid dissociation. If ΔH is equal to ΔH^* no shift is observed, as shown in Eq. (3.1):

$$\Delta H - \Delta H^* = \Delta E_{HA} - \Delta E_A \qquad (3.1)$$

in which ΔE_{HA} is the excitation energy for the 0–0 transition of the acid and ΔE_A is the excitation energy for the 0–0 transition of the base. Assuming the enthropy for the acid dissociation to be the same for the ground state as for the excited state, the following relationship obtains:

$$pK - pK^* = \frac{\Delta E_{HA} - \Delta E_A}{2.3RT} \qquad (3.2)$$

In Eq. (3.2), if $\Delta E_{HA} - \Delta E_A$ is greater than zero, the excited state is more acidic than the ground state as is found in oxy- and amino-aromatic compounds. If $\Delta E_{HA} - \Delta E_A$ is less than zero, the excited state is less acidic than the ground state, as found for acridine and acridone. When $\Delta E_{AH} - \Delta E_A$ is equal to zero, there is no difference in the acidity in the ground state or the excited state as is seen in pyridine.

3.4 Specific Examples of the Effect of pH on Fluorescence

The pH effects have been studied on many systems of compounds, some of which are listed in Tables 3.3, 3.4, and 3.5. The effects on the fluorescence and absorption spectra of these parent compounds with change in pH are given. Also included in Table 3.3 are references for the pH effects upon the fluorescence of derivatives of these parent compounds. The total number of compounds studied with respect to pH dependence on fluorescence has grown so large that these tables are intended to contain only a small but representative sample. The following is a brief description of pH effects on some parent compounds and a short explanation of the origin of the pH effects.

TABLE 3.3 FLUORESCENCE OF CARBAZOLE IN VARIOUS SOLVENTS

Solvent	Fluorescence excitation maxima (nm)				Fluorescence emission maxima (nm)			Ref.
Cyclohexane	290	318	330		332	348		a
Water or 2 N H$_2$SO$_4$	290	323	335		342	355		a
Dimethylformamide	298	339	342		347	360		b
Alkaline dimethylformamide	289	312	383	400	425	450	475	b

[a]Van Duuren, *Anal. Chem.*, **32**, 1436 (1960).
[b]Sawicki et al., *Anal. Chem.*, **33**, 1574 (1961).

TABLE 3.4 FLUORESCENCE OF 2-NAPHTHYLAMINE AT
DIFFERENT pH's[a]

pH	Emission (nm)	Species present[c]
− 5[d]	Ultraviolet[b]	RNH$_3^{\oplus}$
− 1.5[d]	Ultraviolet[b] and 420	RNH$_3^{\oplus}$ and RNH$_2$
2 to 9	420	RNH$_2$
14	420, 530	RNH$_2^{\ominus}$ and RNH$_2$

[a]Excitation at 313 nm, Förster (1960) *Photochemistry in the Liquid and Solid States*.
[b]About 350 nm.
[c]R = 2-naphthyl.
[d]Defined by Hammett and Derup, *J. Am. Chem. Soc.* **54**, 2721 (1932).

One of the most important concepts to remember is that the excited state species of a molecule can have an ionization constant different from that of its ground state, i.e., it can ionize at a different pH than the ground state. If this ionization of the excited state occurs before a photon is emitted, then the emission spectrum may be completely different from that extrapolated from data of the ground state or from spectra at other pH's. For example, even though the ground state has no change in its absorption spectrum there is no reason for this lack of change to be extrapolated to the excited state. This phenomenon is shown in several cases presented below including that of the pyrene trisulfonic acid derivatives, naphthalamine, and the naphthols. Examples of ionization after the molecule is placed in the excited state are given by 3-amino pyrene-1, 6, 8-trisulonic acid and naphthalamine.

The most common effect of pH upon these structures is to protonate or to abstract protons from molecules. If these changes affect the resonance structure of the molecule such as in the case of indole or phenol then the molecular structure responsible for the emission of a photon, i.e., for the $\pi^* \to \pi$, etc., transition, may be lost and the compound then becomes nonfluorescent.

3-Hydroxypyrene-1,6,8-trisulfonic acid (Structure I, R=OH) (Förster, 1950a, 1950b; Peterson, 1949).

I R=OH
II R=O—
III R=OH⁺
IV R=NH₂

The above compound shows an absorption shift to longer wave-lengths above pH 7 which is attributed to the ionization of the phenolic OH yielding structure II. The fluorescence maximum at 510 nm exhibits no concomitant change with pH. As the solution is made more acid, however, from pH 2.0 to pH 0, the 510-nm band decreases in intensity and a new band is observed at 455 nm. Only in very strong acids is the 455-nm band present in the fluorescence spectrum. The 455-nm band is attributed to the protonated form of the phenolic hydroxyl group in the excited state (Structure III), and the 510-nm band to the ionized form of the hydroxyl in the excited state.

3-Aminopyrene-1,6,8-trisulfonic acid (Structure IV, R=NH₂) (Förster, 1950a; Peterson, 1949).

At pH 0 the uv absorption spectrum is similar to pyrene, and above pH 2 up to pH 14 the absorption spectrum becomes that of 3-amino-pyrene. In the pH range of 0 to 12, the fluorescence spectrum remains unchanged with a maximum at 500 nm but shifts to a maximum at 580 nm at pH's from 12 to 14. This change is attributed to the abstraction of a proton from the amine (see reaction scheme below):

$$RNH_2 + h\nu \longrightarrow RNH_2^* \tag{3.3i}$$

$$RNH_2^* + OH^- \rightleftharpoons RNH^{-*} + H_2O \longrightarrow RNH^- + h\nu_{580\,nm} \quad (pH\ 12\text{–}14) \tag{3.3ii}$$

$$RNH_2^* \longrightarrow RNH_2 + h\nu_{500\,nm} \quad (pH\ 12) \tag{3.3iii}$$

It was concluded in the above two cases that ionization equilibrium is reached during the lifetime of the excited state and that the ex-cited states of pyrene compounds are more acidic than the ground state.

TABLE 3.5 EFFECT OF pH ON FLUORESCENCE OF SOME AROMATIC COMPOUNDS

	UV absorption or activation (nm)	Medium	Fluorescence emission maxima, nm	Relative intensity	Ref.
OH	255 264 273[a] 285 316 328	0.2 N H$_2$SO$_4$	358	3.20	b
O$^-$	282 292 346	0.2 N NaOH	429 460	5.90	b
OH OH	288 298 327 337	0.2 N H$_2$SO$_4$	385	0.75	b
O$^-$ OH	321	pH 6.1	453	3.30	b
O$^-$ O$^-$	318 355	0.2 N NaOH	487	4.85	b
NH$_2$	280 280	pH 7–10 pH 14	340 340	1 0	c

Compound		pH			
p-cresol (HO–C₆H₄–CH₃)	285\n285	pH 1–7\npH 14	315\n315	1\n0	c
dimethylphenol (OH, CH₃, CH₃)	285\n285	pH 7\npH 14	310\n310	1\n0	c
COOH–C₆H₄–OH	295\n295	pH 14\npH 1,7	350\n350	1\n0	c
2-hydroxyquinoline	325\n280\n280	pH 1, 7, 10, 14\npH 1,7,10\npH 14	380\n380\n380	1\n1\n1/2	c

^{a} Activation was at any one of these wavelengths, not all at once.
^{b} Hercules and Rogers (1959).
^{c} Williams (1959).

Therefore, dissociation is favored in the excited state. In order to observe the adsorption of the RNH⁻ species (abstraction of a proton from the amine group in the ground state) the solution must be more basic than pH 14. This may be accomplished in liquid ammonia containing sodium amide (Förster, 1952). In such solutions the long wavelength absorption maximum shifts from 415 nm in water and 433 nm in liquid ammonia to 575 nm in liquid ammonia with sodium amide. The latter band is attributed to the RNH⁻ anion.

Indole (Structure I). (Van Duuren, 1961; White, 1959).

Indole in water has a fluorescence maximum at 350 nm. The intensity of the fluorescence is reduced by a factor of 3 in 1 *M* sulfuric acid, the emission maximum remaining at 350 nm. The decrease in intensity has been ascribed to a loss of aromatiticy in the indole cation (Structure II). The above hypothesis is supported by the evidence that in cyclohexane containing trifluoroacetic acid, the relative fluorescence intensity* is $\frac{1}{10}$ that in cyclohexane alone, but the maximum of emission is 297 nm in both cases. There is also a fluorescence quenching above pH 9.5 which is considered to be due to the extraction of a proton to give the indole anion (see Structure III). This is supported by the fact that indole shows no fluorescence in 1 *M* NaOH, whereas *N*-methyl indole fluorescence is unchanged from neutral pH even up to the pH of 5 *M* NaOH solution.

Carbazole (Structure I). (Van Duuren, 1961; Sawicki et al., 1961).

Acidification of carbazole in 1 *M* H_2SO_4 causes no change in the uv absorption spectrum or in the fluorescence spectrum. In alkaline

*The term relative fluorescence intensity has been used by many investigators. Quantum yields and true emission spectra in many of these investigations are not established. However, the results may be considered to be at least semiquantitative. See Chapter 5 concerning quantitative measurements for a more complete description.

N,N-dimethylformamide containing aqueous tetraethyl ammonium hydroxide, however, the uv absorption spectrum and the fluorescence spectrum shift to longer wavelength due to the formation of carbozole anions (Structure II). Table 3.3 gives fluorescence as a function of solvent for carbazole.

2-Naphthylamine (Förster, 1960)

The pK of 2-naphthylamine is 4.07, and the fluorescence emission spectrum with a maximum at 420 nm is unchanged in aqueous solution from pH 2 to 9. The maximum is the same as the fluorescence maximum of 2-naphthylamine dissolved in hexane. Therefore the fluorescence from pH 2–9 in a polar medium eminates from the neutral molecule. The fact that the fluorescence spectrum is that of the neutral molecule even though the solution is acidic enough for the molecule to be protonated in the ground state is explained as due to a loss of a proton from the excited naphthyl ammonium ion (see reaction scheme 3.2). Thus the excited state equilibrium form, *not the ground state form*, determines the fluorescence spectrum:

$$RNH_3^{\oplus} \longrightarrow (RNH_3^{\oplus})* \qquad (3.4i)$$

$$(RNH_3^{\oplus})* \longrightarrow RNH_2^* + H^{\oplus} \qquad (3.4ii)$$

$$(RNH_2)* \longrightarrow RNH_2 + h\nu^1 \qquad (3.4iii)$$

In highly acidic pH solutions such as pH -5 [as defined by Hammett and Deyrup, 1932] only uv emission is observed. This fluorescence is from the naphthyl ammonium ion, whereas at intermediate pH's both the free amine and the naphthyl ammonium ions are present giving two fluorescence bands. At pH 14 the absorption spectrum is unchanged but a new peak in the fluorescence spectrum at 530 nm is observed, the peak at 420 nm disappearing. The change is reversible indicating that there is no change in the essential molecular structure. The shift in fluorescence spectrum is thought to be due to the extraction of a proton from the excited state:

$$(RNH_2)* \xrightarrow{\text{high} \atop \text{pH}} (RNH^{\ominus})* + H^{\oplus} \qquad (3.5)$$

Table 3.4 lists the pH dependence of naphthylamine fluorescence. *N*-methylnaphthylamine shows the same features as the primary amine whereas *N,N* dimethyl-2-naphthylamine does not show any shift in fluorescence spectrum at high pH, which fits the concept of proton extraction in the excited state. The above evidence shows that the naphthylamines are stronger acids in the excited state than in the ground state.

●*Phenol* (White, 1959).

The fluorescence of phenol in alkaline solution follows the titration curve for the phenolic hydroxyl. It is therefore thought that the ionized species (phenolate ion) is nonfluorescent. The fluorescence maximum at pH 7 is 303 nm in water (excitation at 270 nm), but the intensity decreases to zero as the pH is raised to 13. Replacement of the phenolic hydrogen with a methyl (yielding anisole) results in no decrease of the intensity or shift in the emission maximum as the pH is varied from 1–14. The quantum yield over this pH range approximately equals that of phenol at pH 1.

Some studies have been done on the effect of pH on the fluorescence of mono- and dihydroxybenzoic acids and dihydroxybenzene. These compounds show many complications with respect to fluorescence as a function of the number of ionized groups on the benzene ring. A detailed discussion of these results would carry us too far, but the interested reader is advised to consult Table 3.5 for a few cases and Williams (1959) for specific details.

α *and* β *naphthol (1 and 2 naphthol)* (Förster, 1950b, 1960; Hercules and Rogers, 1959; Williams, 1959).

In both naphthols a shift to longer wavelengths occurs in the uv absorption maximum and fluorescence maximum on dissociation of the hydroxyl, the shifts being greater in the emission spectrum than in the absorption spectrum. The shifts are believed to be due to a contribution of the naphthalate ion as shown in structure I, this contribution is from the anionic mesomeric quinoid structures II and III. Thus, fluorescence emission of the undissociated naphthol will be at shorter wavelengths (358 nm) similar to naphthalene. At pH 0 (in 1 *M* HCl) only short wavelength fluorescence maxima are present in 2-naphthol due to the undissociated naphthol. At pH 13 only a long wavelength fluorescence band (460 nm) is present due to the dissociated naphthol. At intermediate pH's both peaks are present. From the work of Hercules and Rogers (1959) and Förster (1950b), the pH's at which the excited and the ground states dissociate have been determined. It is clear that the dissociation occurs at much lower pH's for the excited state than for the ground state, the excited state thus being the stronger acid. The remarkably wide pH range at which both naphthol and naphtholate are present as shown by the fluorescence spectrum has been interpreted by Förster as evidence of incomplete ionization equilibrium established during the lifetime of the excited state. Figure 3.8 [taken from Hercules and Rogers (1959)] shows the

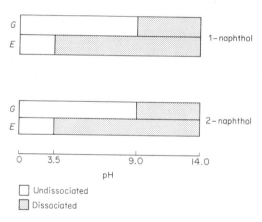

Fig. 3.8. Ground (*G*) and excited state (*E*) dissociation of 1- and 2-naphthol (Hercules and Rogers, 1959).

pH ranges from which both 1 and 2 naphthol exist as dissociated or associated forms in the ground and excited states. (See Table 3.5 for a few examples of naphthols.)

REFERENCES

Bayliss, N. S., and Mehar, E. G. (1954). *J. Phys. Chem.*, **58**, 1002.
Brealey, G. J., and Kasha, M. (1955). *J. Am. Chem. Soc.*, **77**, 4462.
Förster, T. (1950a) *Z. Elektrochem.*, **54**, 42.
Förster, T. (1950b). *Z. Elektrochem.*, **54**, 531.
Förster, T. (1952). *Z. Elektrochem.*, **56**, 716.

Förster, T. (1960). *Photochemistry in the Liquid and Solid States* (L. J. Heidt, R. S. Livingston, E. Rabinowitch, and F. Daniels, eds.), Wiley-Interscience, New York and London, p. 10.

Hammett, L. P., and Deyrup, A. J. (1932). *J. Am. Chem. Soc.*, **54**, 2721.

Hercules, D. M., and Rogers, L. B. (1959). *Spectrochim. Acta*, **14**, 393.

Matage, N., and Tsuno, S. (1957). *Bull. Chem. Soc. Japan*, **30**, 368.

McConnel, H. (1952). *J. Chem. Phys.*, **20**, 700.

Peterson, S. (1949). *Angew. Chem.*, **61**, 17.

Pimentel, G. C. (1957). *J. Am. Chem. Soc.*, **79**, 3323.

Sawicki, E., Hauser, T. R., Stanley, T. W., Elkert, W., and Fox, F. T. (1961). *Ann. Chem.*, **33**, 1574.

Van Duuren, B. L. (1961). *J. Org. Chem.*, **26**, 2954.

Van Duuren, B. L. (1963). *Chem. Rev.*, **63**, 325.

Weller, A. (1957). *Z. Elektrochem.*, **61**, 956.

White, A. (1959). *Biochem. J.*, **71**, 217.

Williams, R. T. (1959). *J. Roy. Inst. Chem.*, **83**, 611.

Chapter 4

Polarized Fluorescence

4.1 Introduction

A. POLARIZATION

The wave model of light is depicted in Fig. 4.1 with its mutually perpendicular velocity, electric field, and magnetic field vectors represented by **V**, **E**, and **H**, respectively. This single ray may be characterized by the orientation of its electric vector within a coordinate system. Thus, if one looked directly into an oncoming ray, the electric

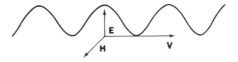

Fig. 4.1

vector would be at some angle θ, as shown in Fig. 4.2a. Likewise, a beam composed of a number of rays could be characterized by the angular distribution of its electric vectors. Two extreme situations are possible: the vectors can be evenly (randomly) distributed as in Fig. 4.2b, or they can be parallel as in Fig. 4.2c and 4.2d. Experimentally, these two situations can be differentiated by viewing the light through a polarizer which allows maximum transmission of light when it is aligned parallel to the electric vector and zero transmission when it is perpendicular. Therefore, as the polarizer is turned through 180°*

---

*Electric vectors have no heads or tails, so if the polarizer were turned the remaining way from 180° to 360° the result would be equivalent.

87

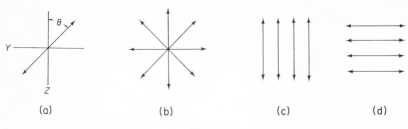

Fig. 4.2

in Fig. 4.2b, the intensity remains constant. However, in Fig. 4.2c the intensity would be a maximum with the polarizer vertical and zero with it horizontal, the converse being true for Fig. 4.2d. It is this experimental observation which allows one to define the polarization of light as

$$p = \frac{I_{\parallel} - I_{\perp}}{I_{\parallel} + I_{\perp}} \tag{4.1}$$

where $I_{\parallel}$ and $I_{\perp}$ refer to the vertical and horizontal intensity components of the light relative to the observer.

Hence, in Fig. 4.2b, the two components are always equal and $p = 0$. In Fig. 4.2c however, $I_{\perp}$ is 0 and $p = 1$, while in Fig. 4.2d, $I_{\parallel} = 0$ and $p = -1$. When $p = 0$, the light is called unpolarized or natural light, with $p = \pm 1$, it is completely polarized, whereas with $1 \geqslant p \geqslant -1$, but not equal to 0, it is partially polarized.

B. DIPOLE MODEL OF ABSORPTION AND EMISSION

Figure 4.3 depicts the experimental arrangement whereby most fluorescence parameters are measured. If a polarizer is placed at C, and the fluorescence intensity components parallel ($I_{\parallel}$) and perpendicular ($I_{\perp}$) to the Z-axis measured, the polarization of the fluorescent light may be determined. It is found that certain fluorescent solutions can yield partially polarized light when they are excited with either *unpolarized* or *completely polarized* light.* We may ask how it is that in the latter case the complete polarization of the exciting light is lost upon absorption and reemission, while in the former, the complete lack of polarization is lost. To answer this question, we must first consider several aspects of the interaction of light and matter.

---

*In our discussions, polarized exciting light will always have its electric vector parallel to the Z axis, while unpolarized light will, of course, have its vectors anywhere in the YZ plane. When we speak of the orientation of an emission oscillator this is the angle its transition moment of emission makes with the Z axis.

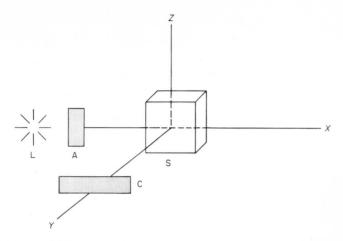

Fig. 4.3. Experimental arrangements for measuring fluorescence polarization. L = light source; A and C = various electronic and optical components; S = sample; X, Y, and Z = axes.

Light is absorbed by and emitted from molecules via electric dipole oscillators which have certain orientations relative to the structure of the molecule. In a small molecule which has a considerable overall nuclear ridigity, the dipole has a definite orientation relative to the entire molecule (Fig. 4.4a). In a larger molecule with greater flexibility (i.e., a polymer) the dipole may have only a time-averaged orientation to the entire structure (Fig. 4.4b). In either event, a random distribution of molecules results in a random distribution of dipoles.

Maximum absorption of light occurs when the absorption dipole is parallel to the electric vector of the light. Specifically, the probability of absorption is proportional to $\cos^2 \theta$ where θ is the angle between the dipole and the electric vector. Therefore, absorption is a maximum when $\theta = 0°$ and is zero when $\theta = 90°$.

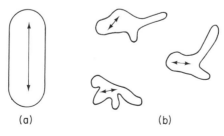

(a)　　　　　(b)

Fig. 4.4

Emission from a dipole can occur with the electric vector at any angle to the dipole, but the probability is proportional to $\sin^2 E$ where E is the angle between the dipole and the electric vector. As the number of dipoles that emit at a given angle is proportional to the probability of emission at that angle, the intensity of the light must also be proportional to $\sin^2 E$.

With these few statements, we can now give a qualitative picture of how polarized fluorescence arises and the factors which influence it.

C. Qualitative Aspects of Fluorescence Polarization

The sequence of events in fluorescence is depicted in Fig. 4.5 and is briefly recapitulated as follows: (1) A quantum of light is absorbed

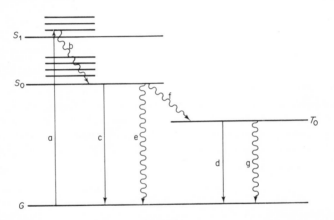

Fig. 4.5

by a dipole (a). (2) Rapid radiationless relaxation takes the dipole to the thermally relaxed S_0 level (b). (3) The dipole retains this energy for a certain period of time, generally designated as the lifetime τ. (4) The excess energy is dissipated through radiative (c,d) and/or nonradiative (e,f,g) processes. In each of these processes, changes can occur such that the polarization of the fluorescent light is different from that of the exciting light. Let us illustrate this by considering a two-dimensional system (in the plane of paper) of random molecules (Fig. 4.6a) undergoing excitation with *vertically polarized* light, with the fluorescence being observed perpendicular to the plane of the paper. From previous remarks, we know that the vertical dipoles will be preferentially excited to give an excited dipole distribution similar to Fig. 4.6b. Emission from these dipoles would give only partial

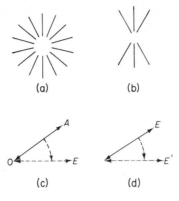

Fig. 4.6

polarization, as nonvertical dipoles do absorb some of the vertically polarized exciting light. Therefore, the absorption process itself is the first source of depolarization. (Similarly, excitation with unpolarized light yields an uneven distribution; in this case, the process of absorption is an act of polarization.)

Now consider just one of these excited absorption dipoles, OA. If the emission dipole OE is at an angle to OA, as in Fig. 4.6c, a second source of depolarization is evident which is dependent upon the angle between the two dipoles, the larger the angle the greater the depolarization. The electric vector of the light can be said to have undergone two

TABLE 4.1

Process	Effect on Polarization
Absorption	Selection of excitable dipoles, polarization
Intramolecular transfer between absorption and emission oscillators	Change in orientation of dipole, intramolecular depolarization
Brownian motion	Rotation of excited dipole, depolarization
Energy transfer	Loss of orientation by transfer to nonparallel dipoles, depolarization
Reabsorption of emitted light	Loss of orientation by transfer to nonparallel dipoles, depolarization
Light scattering	Loss of orientation of both exciting and emitted light, depolarization

displacements, thus far. The first one is from the vertical position to that of the absorbing dipole, and the second is from the absorbing dipole to the emitting dipole.

A third depolarizing factor is the Brownian rotation of the molecule. As the emitting dipole has some geometric orientation relative to the molecular structure (static or statistical), rotation of the molecule will result in rotation of the dipole OE to a new orientation OE′, as represented in Fig. 4.6d. Therefore, *if* the excited state lifetime is comparable to the time required for Brownian rotation to occur over an appreciable angle, significant depolarization results. The extent of this depolarization is determined primarily by the following factors: (1) size and shape of the unit containing the dipole, (2) temperature, (3) viscosity, and (4) excited state lifetime.

Figure 4.6d can also depict the depolarization resulting from nonradiative energy transfer. As discussed in Chapter 5, transfer takes place preferentially between parallel dipoles, but nonparallel dipoles also undergo transfer with resulting depolarization. The extent of this depolarization is dependent upon the spectral properties of the absorbing and fluorescing species and the concentration.

A final mechanism of depolarization is trivial reabsorption: the light emitted from one molecule is absorbed by another molecule which then reemits it at a different orientation. The extent of this depolarization depends again on the spectral properties of the molecule, the concentration, and the average light path. From the relationships of these factors, fluorescence polarization can be used to investigate both

TABLE 4.2

Property	Conditions
1. Angle between absorption and emission dipoles	No Brownian motion, trivial reabsorption, or energy transfer; therefore, high viscosity and/or low temperature and low concentration.
2. Energy transfer characteristics	High viscosity and high concentration
3. Molecular size	Changing viscosity
4. Conformational changes	Changing temperature and/or environment
5. Binding of small molecules to large molecules	Small fluorescent molecule or significant change of large molecule upon binding.

spectral and molecular properties. Table 4.1 summarizes the factors and their effect on polarization while Table 4.2 summarizes the meaurable properties and conditions required for their measurement.

With this introduction, we shall now discuss the quantitative relationships of fluorescence polarization and conclude with a survey of its applications.

4.2 Quantitative Aspects of Fluorescence Polarization

The degree of polarization of fluorescence is a function of the geometrical orientation of the exciting light, absorbing dipole, emitting dipole, and changes of dipole orientation while in the excited state. The effect of each of these factors on polarization is given in Table 4.3. The derivations of each of these equations and the limits of the values of p are given in the appendix. The final form of the equation is numbered to give a cross reference to the corresponding derivation. One geometrical consideration which simplifies the calculations is the Soleillet equation which shows that the change in polarization for any additional change in orientation will be accounted for by the factor $2/(3 \overline{\cos^2 \beta} - 1)$. To establish the exact value of the factor, the physical boundaries and relationships must then be considered. In cases 7 and 8 of Table 4.3, the factor becomes $(1 + RT\tau/V\eta)$, and for energy transfer (case 9) it becomes $(1 + \frac{3}{2} \overline{\sin^2 \theta_A \bar{n}})$.

The term p_0 is defined as the intrinsic polarization of the molecule in the absence of any depolarization due to Brownian motion or intermolecular processes. It is a function only of the polarization due to absorption of light and of the intramolecular angle of transfer between the absorption and emission dipoles. (See also Table 4.4 for an operational definition of p_0).

The range of values for p_0 when a molecule is excited by polarized light is

$$-\frac{1}{3} \leqslant p_0 \leqslant \frac{1}{2}.$$

When excited with unpolarized light the range of values is

$$-\frac{1}{7} \leqslant p_0 \leqslant \frac{1}{3}.$$

Thus the limiting values of p_0 measured for any systems should be -0.33 and $+0.500$.

TABLE 4.3 Summary of Polarization Equations

Physical orientations	Equation	Dipole[a]/Orientation		Value of p_0
General equations				
1. Exciting light polarized	$\dfrac{1}{p} - \dfrac{1}{3} = \dfrac{\frac{2}{3}}{\frac{1}{2}(3\cos^2\theta - 1)}$ (A4.7)	E/Random	$\overline{\cos^2\theta} = \dfrac{1}{3}$	0
		E/Parallel to x axis	$\overline{\cos^2\theta} = 1$	1
2. Exciting light unpolarized	$\dfrac{1}{p} + \dfrac{1}{3} = \dfrac{-\frac{2}{3}}{\frac{1}{2}(3\cos^2\theta - 1)}$ (A4.19)	E/Random	$\overline{\cos^2\theta} = \dfrac{1}{3}$	0
		E/Parallel to y axis	$\overline{\cos^2\theta} = 1$	-1
Immobile parallel absorption and emission oscillators				
3. Exciting light polarized	$\dfrac{1}{p} - \dfrac{1}{3} = \dfrac{\frac{2}{3}}{\frac{1}{2}(3\cos^2\theta - 1)}$ (A4.7)	A/Random	$\overline{\cos^2\theta} = \dfrac{3}{5}$	$\dfrac{1}{2}$
4. Exciting light unpolarized	$\dfrac{1}{p} + \dfrac{1}{3} = \dfrac{-\frac{2}{3}}{\frac{1}{2}(3\cos^2\theta - 1)}$ (A4.19)	A/Random	$\overline{\cos^2\theta} = \dfrac{1}{5}$	$\dfrac{1}{3}$
Immobile nonparallel absorption and emission oscillators				
5. Exciting light polarized	$\dfrac{1}{p_0} - \dfrac{1}{3} = \dfrac{5}{3}\dfrac{2}{3\cos^2\lambda - 1}$ (A4.39)	$A + E$/Parallel to one another	$\lambda = 0$	$\dfrac{1}{2}$
		$A + E$/Perpendicular to one another	$\lambda = \dfrac{\pi}{2}$	$-\dfrac{1}{3}$
6. Exciting light unpolarized	$\dfrac{1}{p_0} + \dfrac{1}{3} = \dfrac{20}{3}\dfrac{1}{3\cos^2\lambda - 1}$ (A4.43)	$A + E$/Parallel to one another	$\lambda = 0$	$\dfrac{1}{3}$
		$A + E$/Perpendicular to one another	$\lambda = \dfrac{\pi}{2}$	$-\dfrac{1}{7}$

Mobile nonparallel absorption and emission oscillators
Brownian motion

7. Exciting light polarized	$$\frac{1}{p} - \frac{1}{3} = \left(\frac{1}{p_0} - \frac{1}{3}\right)\left(\frac{2}{3\,\overline{\cos^2\beta} - 1}\right)$$ (A4.46)	E/Parallel[b] (no displacement)	$\overline{\cos^2\beta} = 0$	$p = p_0$
		E/Random[b] (complete displacement)	$\overline{\cos^2\beta} = \frac{1}{2}$	$p = 0$
8. Exciting light unpolarized	$$\frac{1}{p} + \frac{1}{3} = \left(\frac{1}{p_0} + \frac{1}{3}\right)\left(\frac{2}{3\,\overline{\cos^2\beta} - 1}\right)$$	E/Parallel[b] (no displacement)	$\overline{\cos^2\beta} = 0$	$p = p_0$
Energy transfer		E/Random[b]	$\overline{\cos^2\beta} = \frac{1}{2}$	$p = 0$
9. Exciting light polarized	$$\left(\frac{1}{p} - \frac{1}{3}\right) = \left(\frac{1}{p_0} - \frac{1}{3}\right)\left(1 + \tfrac{3}{2}\,\overline{\sin^2\theta_A\bar{n}}\right)$$ (A4.84)	E/Parallel[b]	$\overline{\sin^2\theta_A} = 0$	$p = p_0$
		E/Random[b]		$p = 0$

[a] A = Absorption dipole. E = Emitting dipole.
[b] Refers to orientation between emitting dipoles before and after migration by either Brownian motion or energy transfer.

4.3 Application of Fluorescence Depolarization to Determine the Structure of Molecules

A. PERRIN'S EQUATION

A theory for the quantitative relationships between p, τ, and ρ was formulated by Perrin (1926) and was expanded to cover biological applications by Weber (1952a). A general theory covering molecules of various shapes and having various relative orientations of absorption and emission oscillators has been presented by Memming (1961). The basic equation for a sphere is:

$$\frac{1}{p} \pm \frac{1}{3} = \left(\frac{1}{p_0} \pm \frac{1}{3}\right)\left(1 + \frac{\tau}{\rho}\right) = \left(\frac{1}{p_0} \pm \frac{1}{3}\right)\left(1 + \frac{RT\tau}{\eta V_0}\right) \tag{4.2}$$

where p = polarization
$\quad\quad p_0$ = limiting or intrinsic polarization
$\quad\quad \tau$ = excited state lifetime
$\quad\quad \rho$ = rotational relaxation time
$\quad\quad R$ = universal gas constant
$\quad\quad T$ = temperature
$\quad\quad \eta$ = viscosity
$\quad\quad V_0$ = volume of equivalent sphere

The positive signs apply to excitation with natural (unpolarized) light and the minus signs to excitation with polarized light.

The assumptions for Eq. (4.27) are: (1) The rotating particle is a sphere of volume V or has the same rotational properties as a sphere of volume V_0. (2) The medium is continuous. (3) Microviscosity equals the solution viscosity. (4) Depolarization occurs as a result of the difference in orientation between the absorption and emission oscillators (p_0) and by Brownian motion. (5) Rotation is random. (6) There is only one kind of fluorescent molecule. It will become clear that these assumptions will not hold for all experimental cases.

The predictions which can be made from Eq. (4.1) are:

(1) $1/p \propto T/\eta$, with intercept $1/p_0$

(2) $1/p \propto 1/\rho$ and ideally $1/V_0$

(3) $1/p \propto \tau$

Relationship (1) has been verified with numerous systems for which plots of $1/p$ vs T/η were linear and yielded p_0 values equal to p values obtained in viscous media such as glycerol or 60% sucrose where negligible rotation should occur.* However, there are also cases where a straight line is not obtained. Both of these cases will be discussed in greater detail later. There is less experimental evidence for (2) and (3). The best rationale for their validity is that, in a number of cases, values of ρ or τ determined independently have agreed with the results of polarization measurements.

The simplifying assumption essentially excludes energy transfer between nonparallel, nonperpendicular oscillators which also leads to depolarization. In summary, depolarization can arise according to the scheme shown in Table 4.4 according to Weber (1966).

TABLE 4.4[a]

Condition	Observed polarization	Molecular properties responsible
1. Dilute solution in very viscous solvents	Maximum possible p_0 for given wavelength	Electronic structure of molecule
2. Concentrated solutions in very viscous solvents	Decrease in polarization due to energy migration	Electromagnetic coupling of neighboring molecules
3. Dilute solutions in solvents of relatively low viscosity	Decrease in polarization due to Brownian motion	Size and shape of the kinetic unit

[a]Taken from Weber (1966).

B. WEBER'S EQUATIONS

Weber (1952a) has extended the work of Perrin to include collections of particles, ellipsoids, and applications to macromolecules. For a collection of different oscillators (molecules which have their absorption and emission dipoles at all possible orientations), the average polarization is the harmonic mean of the individual p's:

$$\frac{1}{\bar{p}} = \frac{1}{p_1} + \frac{1}{p_2} + \cdots \cdot \frac{1}{p_n} \tag{4.3}$$

If the case of two oscillators is considered, a plot of $1/p$ vs T/η (ρ plot) will yield a straight line if τ_1/τ_2 is less than $\frac{3}{2}$, but a line concave to the T/η axis if it is greater than $\frac{3}{2}$. If one has the same oscillator, but a collection of different sized particles, a ρ plot will yield at low

*(Note: a plot of $1/p$ vs T/η will hereafter be called a ρ plot.)

values of T/η an initial slope which is proportional to the harmonic mean of the ρ's:

$$\frac{1}{\rho_h} = \frac{1}{\rho_1} + \frac{1}{\rho_2} + \frac{1}{\rho_3} + \cdots + \frac{1}{\rho_n} \qquad (4.4)$$

At higher values of T/η there will be curvature when ρ_1/ρ_2 is greater than $\frac{2}{3}$. Weber and Teale (1965) pointed out that varying T/η solely by changing the temperature may possibly break bonds which can create or increase internal rotation or dissociation; therefore, the polarization measurements may not represent the whole molecule or even the same unit throughout the range of the experiment (Fig. 4.9). One can, however, change T/η at constant temperature by adding to the solvent substances such as sucrose or glycerol which will increase the viscosity. If internal rotations are created, a plot of $1/p$ vs T/η can be convex to the T/η axis. Hence, the predicted plots one may obtain are depicted in Fig. 4.7.

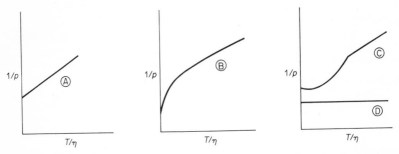

Fig. 4.7. Possible results of plots of T/η vs $1/p$. A: Normal. B: More than one rotational relaxation time or lifetime. C: Creation of rotating units of smaller equivalent volume at high temperature. D: ρ too large or τ too small to make possible the measurement of changes in polarization.

The simplicity of the Perrin equation plus its validity in many situations had deterred many investigators from utilizing an equation derived by Weber (1952a, 1953) for ellipsoids of revolution:

$$\frac{1}{p} - \frac{1}{3} = \frac{\left(\frac{1}{p_0} - \frac{1}{3}\right)\left(1 + \frac{3\tau_0}{\eta_1\rho_0}\right)\left(1 + \left[\frac{4}{\eta_2} - \frac{1}{\eta_1}\right]\frac{\tau_0}{\rho_0}\right)}{1 + \left(\frac{2}{\eta_2} + \frac{1}{\eta_1}\right)\frac{\tau_0}{\rho_0} - \frac{6}{5}(\tau_0\rho_0)^2 \left(\frac{1}{\eta_2} - \frac{1}{\eta_1}\right)^2 \Big/ \left(1 + \left[\frac{2}{\eta_1} + \frac{1}{\eta_2}\right]\frac{\tau_0}{\rho_0}\right)} \qquad (4.5)$$

where ρ_1 and ρ_2 are the relaxation times around the long and short axes, respectively and

$$\eta_1 = \frac{\rho_1}{\rho_0}; \qquad \eta_2 = \frac{\rho_2}{\rho_0}; \qquad \rho_0 = \frac{3\eta V_0}{RT}$$

The assumptions are the same as those for the Perrin equation except that it applies to an ellipsoid rather than a sphere, and the oscillators are assumed to be oriented randomly with respect to the axes of the ellipsoid. It is seen that, when $\eta_1 = \eta_2 = 1$, the equation reduces to the simpler Perrin equation.

C. PROTEIN CONJUGATES

Realizing that most macromolecules either do not have fluorescent groups or have groups whose excited lifetimes are too short in comparison with the rotational relaxation times, Weber (1952b) introduced the use of covalently linked protein-dye conjugates for polarization studies. If it is assumed that the dye-protein conjugate is a rigid particle, the rotational properties of the dye are identical to those of the macromolecule. In cases where one or more of the dye molecules attached to a given particle has its own freedom of rotation, the ρ plot can show two relaxation times. In addition to covalently linked dyes, it is possible to utilize the ability of macromolecules to adsorb certain dyes (Laurence, 1952). That situation is somewhat more complex to analyze since there is both unbound (free) and bound dye for determination of a protein structure by means of a conjugate. The criteria for using fluorescence conjugates are (Weber, 1952b):

(1) minimum change of protein upon coupling;
(2) only one type of bond between dye and protein to ensure a single lifetime;
(3) fluorescence efficiency of coupled dye not appreciably smaller and preferably much larger than that of free dye to assure reduced contamination of fluorescence of free dye;
(4) stability of dye-protein unaltered compared to the stability of unconjugated protein;
(5) fluorescence intensity of the dye-protein complex constant over a wide range of conditions.

These criteria are identical for all macromolecule, but proteins have formed the vast majority of the conjugates studied thus far.

In spite of the above practical problems, the advantages of polarization of fluorescence over some of the other hydrodynamic methods include the following:

(1) applicability at low concentrations of protein;
(2) rapid analysis (equilibrium of solution is virtually instantaneous);
(3) allows wide variation of ionic strength, solutes, and pH;
(4) highly sensitive to certain small changes; e.g., internal rotations.

D. Experimental Summary

Experimentally, p is determined as a function of some variable. An analysis of Eq. (4.27) indicates that the following relationships may be utilized, each of which can also be experimentally realized:

(1) $1/p$ vs T/η yields p_0 from the intercept and either V_0 or ρ from the slope, if τ is known. (T/η can be varied by changing the temperature, the viscosity, or both.)

(2) $1/p$ vs T/η at known ρ yields τ.

(3) $1/p$ vs "parameter," where the latter is something which can affect the system, indicates changes in one or a combination of the following: temperature, viscosity, τ, V_0, ρ, or p_0.

(4) p vs λ (excitation) yields p_0 values if the measurement is done in a highly viscous solution; these can then be used to analyze absorption bands.

The applications of each of these treatments will be briefly discussed drawing upon examples from the literature (see Section E).

As the data for a ρ plot require a multistep experiment, it is practical to determine the conditions yielding the optimal information. These conditions are usually determined by one of two ways. First, they can correspond to conditions utilized in an independent experiment, which allows a direct comparison of the results. Second, one can vary the conditions, observe the changes or constancy of p, and select those conditions which seem fruitful for investigation. Due to the large application of this approach, it will be useful to consider the two experimental treatments of ρ plots and parameter plots simultaneously.

The parameters which can be varied include the following: (1) pH; (2) concentration of denaturing agents such as urea, guanidine, dodecyl sulfate, acid or base; (3) concentration of various salts, small molecules such as cofactors, substrates, haptens, allosteric effectors, and quenchers; (4) concentration of the fluorescent material itself, and of various other reactive species; (5) temperature; and (6) time. Such plots of p or $1/p$ versus any of these variables will be termed a parameter plot. The assumptions employed are that all factors other than the parameters p, and V_0 or ρ are constant. Because this is not always valid, it is necessary to experimentally obtain a ρ plot at each set of conditions of interest. Two factors which would go unnoticed were one to calculate ρ from single values of p are (1) changes of p_0 and (2) nonlinear ρ plots. Either of these would invalidate ρ values. Changes in p_0 and nonlinear plots of $1/p$ vs T/η_0 are useful information on the state of the system. It cannot be overemphasized that single values of p, in themselves, are not sufficient for determination of ρ values.

E. Specific Examples of the use of Depolarization of Fluorescence

Singleterry and Weinberger (1951) introduced the use of fluorescence polarization for the determination of macroparticle volumes when they investigated soap micelles labeled with adsorbed Rhodamine B in organic solvents, e.g., calcium xenylstearate in benzene. This method proved to be more rapid and sensitive than osmometry, light scattering, or viscometry, and gave comparable values for gram-micellar weights.

The principal use of this technique, however, has been with proteins. The main reasons for this are: (1) Macromolecular properties of proteins are so widely studied by a number of independent techniques that correlations and comparisons are readily available. (2) Labeling agents have been developed which are versatile and dependable. (3) Conformational changes have been proven to be of great significance to their function. (4) Their rotational diffusion properties are such as to make them susceptible to this technique. (5) Interactions between proteins and other molecules constitute the foundation for many biological reactions. Therefore, examples of this particular application will be divided into four different areas.

1. *Structural Changes of Macromolecules*

The significant extension of fluorescence polarization to macromolecules is due to Weber's theoretical formulation (1952a) and initial studies (1952b) with dimethylaminonaphthalene sulfonic acid (DNS) conjugates of bovine serum albumin (BSA) and ovalbumin (BSA-DNS, Ovalbumin-DNS). The denaturation by acid, base, and urea of both conjugates was observed. Whereas the ρ value of the denatured ovalbumin is larger than that of the native molecule, the ρ value of acid denatured BSA appeared to be smaller. The denaturation of ovalbumin was explained by aggregation, whereas that of serum albumin was originally explained by fission into subunits, which were later assumed to be linked by peptide chains (Weber & Young, 1964). Assuming the ρ value of ovalbumin as determined by Oncley (75.5 nsec), τ was calculated to be 14 nsec. By using this value, ρ for BSA was determined by a ρ plot to be 127 nsec, which compares well with the value determined by Oncley (1942).

The use of adsorbed as opposed to covalently linked dyes was championed by Laurence (1952), and its validity in ρ plots was tested with DNS adsorbed to BSA. The slope was half that of the conjugate and the intercepts equal. According to Weber (unpublished results), the lifetime of DNS in glycerol is about twice that in water. Hence, assuming the lifetime of DNS in glycerol is about equal to that of the adsorbed DNS, the ρ values are equivalent.

Steiner and McAlister (1957) studied various conjugates of lysozyme, BSA, and ovalbumin using fluorescein, anthracene, and DNS as the dyes. A valuable part of this work was the determination of the lifetime of each conjugate at a series of pH values. The lifetime of the excited state of fluorescein was found to be quite constant between pH 4.5 and 10 with an approximate value of 4.8 nsec. Similar stability was found for the DNS derivatives, which had a lifetime of 12.5 nsec, agreeing relatively well with the value of Weber (1952b). There was considerable variability, however, in the anthracene conjugates of ovalbumin and BSA. Given a particular conjugate of anthracene, however, τ was found to be independent of pH and ionic strength. In general, they found the apparent value of ρ for any of the proteins was essentially independent of the nature of the conjugate, the degree of conjugation, the value of τ, and the exciting wavelength. This constancy, in spite of the diverseness of lifetimes p_0 values and reaction sites, offers further evidence for the assumption of random distribution of the oscillators (Weber and Teale, 1965)

2. Interaction of Macromolecules

An increasing use of fluorescence polarization is for the study of the interactions of antibodies with antigens or haptens. The main principle is that the ρ value of the complex is larger than the ρ value of the originally fluorescent unit. Hence, the greatest accuracy is obtained when the difference between the two ρ's is the largest possible. One therefore labels either the hapten or the smaller of the two proteins. This technique has found special applicability to soluble complexes which cannot be analyzed by the usual immunochemical techniques.

Dandliker and co-workers (1965) have formulated a theory for the interpretation of polarization measurements as applied to molecular interactions. A titration technique is used with the fluorescent dye as the titrant. The curves obtained of polarization versus the concentration of labeled material are a function of both the state of aggregation and the relative proportions of bound and free label. Analysis of the curves yields information about the association constant, concentration of binding sites, and some measure of the heterogeneity of the association constant. The system was tested by using fluorescein-labeled ovalbumin and its antibody. One of the principal advantages of this technique is the small amount of material required for analysis.

3. *Interaction of Small Molecules with Macromolecules*

Laurence (1952), in the study of equilibrium relationships between large molecules and adsorbed dyes by the use of fluorescence polarization, used dyes of two types. One had essentially the same quantum yield in the free and adsorbed states, while the other had a significant quantum yield only in the adsorbed state. The value of this second type of system is readily appreciated (see Chapter 7).

4. *Degradation and Polymerization Reactions*

The use of polarization studies for polymerization and degradation studies is obvious. One of the first applications was by Arkin and Singleterry (1949) when they studied the effect of increasing amounts of water on the micelle size of calcium xenylstearate in benzene with the aid of an adsorbed dye. The polarization curves correlated well with intrinsic viscosity curves. At high intrinsic viscosity, the polarization data suggested that the micellar structure was unable to undergo significant Brownian motion. As water was added, the viscosity decreased, and the polarization decreased until the predominant specie was a monomer unit.

F. ρ PLOT YIELDING p_0 AND τ

The application of this relationship is valid only if the values of ρ or V_0 of the rotating species are known. The most common application of this relationship has been the determination of the lifetime of a fluorescent group which is covalently bound or adsorbed to a rigid molecule whose rotational relaxation time ρ is known, for example, the lifetime of DNS bound to ovalbumin. If this dye is to be utilized in other systems, the lifetime must be ascertained to be the same. This constancy of lifetime can be indirectly determined through quantum yield measurements, careful analysis of the absorption and emission spectra, and determination of p_0 values. However, some uncertainty in the value of τ still remains.

G. ρ AS AN INDICATION OF A CHANGE OF VISCOSITY, CHANGE OF τ

Assuming all other parameters are constant, changes in p can be used to monitor changes in viscosity. This approach was utilized by Buchader and Lebesque (1946) in a study of the transformation of a sodium silicate solution of sulfuric acid into a gel. Rhodamine S was added to the initial solution. With increased gel formation, polarization and quantum yield increased. The authors suggested that these facts were contradictory since increased quantum yield should give increased lifetime which results in decreased polarization. (This is not necessarily true; this point will be discussed in more detail later). However, the authors concluded that the overall effect was not due to the balance of increased depolarization through an increased lifetime, but to increased polarization through adsorption of the dye to the colloidal particles being formed. It is possible that the increased polarization could also have been due to the increased microviscosity of the medium.

H. p VERSUS WAVELENGTH OF EXCITATION, YIELDING $p_0(\lambda)$

If a molecule is studied in a solution of sufficiently high viscosity or low temperature, then the Brownian motion may be negligible. In this case, the only depolarization will be from the intramolecular depolarization ($p_0 < 1$) and energy transfer. Assuming that energy transfer is negligible, it is possible to analyze absorption spectra through a plot of p_0 vs wavelength of excitation. If the polarization remains constant with wavelength, it can be assumed that only one absorption band is present. If the absorption transition changes with wavelength (the emission dipole in almost all cases remains the same, cf. Chapter 2), the angle between the two dipoles changes and p_0 changes. Therefore, changes in the polarization occur as a function of wavelength. Whereas a given absorption band may appear to be singular in an ordinary spectrophotometer, it may actually consist in a combination of two or more transitions (see Fig. 4.8).

By careful study of the polarization spectrum of a compound it is possible to enumerate the number of absorption bands in a molecule. Furthermore, it may be possible to determine perturbation effects on

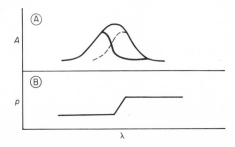

Fig. 4.8. A: Absorption spectrum. B:Fluorescence polarization spectrum.

the individual bands which might be indistinguishable from a broad absorption spectrum and it may be possible to distinguish between vibrational structures and the overlap of electronic bands.

I. COMPLICATIONS AND LIMITATIONS

1. ρ plots

a. *Interpretations of ρ and V_0 Values.* The basic Perrin equation assumes the rotating particle to be a sphere. Therefore, the value determined is for the equivalent sphere, i.e., that sphere which would give the same mean rotational relaxation time as the molecule being studied. In many cases this may be close to that of the actual protein, but often it is not. Another problem is the water of hydration which varies with each protein and could introduce significant error in the interpreted meaning of ρ or V_0, even if the shape were spherical. If the particle is actually an ellipsoid or rod-shaped, then a number of axial ratios could lead to the same V_0 or ρ. In addition, the rotating particle may not even be the entire molecule, but may be some fragment which is able to randomly orient itself (Kuhn segment), a more restricted fragment, or the dye molecule itself (Figs. 4.9 and 4.10). In cases where both a local rotation and a more extensive rotation can make significant

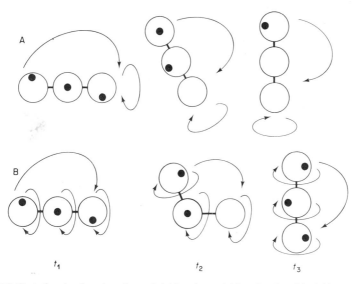

t_1 t_2 t_3

Fig. 4.9. Rotational relaxation time of rigid and nonrigid molecule with rigid segments. A: rigid molecule rotates as one unit. B: model rotates so that both long and short axis rotations are not the same as the whole molecule.

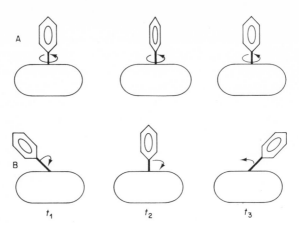

Fig. 4.10. Modes of rotation of a dye on a rigid molecule, dye is enlarged for illustration. A: rotation of dye around linkage. B: "binding" rotation of dye on molecule.

contributions, the net ρ value is the harmonic mean of the independent ρ values (Weber, 1952a). Caution is necessary when considering interpretation of ρ values and V_0 values. The final interpretation must await support from other hydrodynamic methods (flow birefringence, diffusion, sedimentation, viscosity) and X-ray crystallography. In this way obvious misinterpretations may be prevented (Weber, 1952a). In the case of proteins whose structure is known, the interpretation of fluorescence polarization can be meaningful and contribute to a further testing of the theories concerning spheres, ellipsoids, and rods. This correlation has been applied in a number of studies with varying degrees of success (Weber, 1952b; Churchich, 1963).

    b. *Nonlinear ρ Plots.*   Another major problem is the interpretation of nonlinear plots (Fig. 4.7). It is assumed that curvature concave to the T/η axis is due to the presence of more than one relaxation time which may either be the ρ values of an ellipsoid or of the freely rotating dye plus part or all of the macromolecule. The distinction between these two cases is not easy to make without independent data for axial ratio values. Curvature convex to the T/η axis is attributed to (1) the creation of internal degrees of rotation or (2) division of the molecule into subunits. Both of these situations can arise by changing the temperature, but can be distinguished by dilution, which facilitates dissociation but not internal rotational freedom. The effect, itself, can usually be eliminated by changing the viscosity instead of the temperature. However, it is a good idea to check each ρ plot both ways, if possible, to avoid specific artifacts arising from temperature effects or the influence of the sucrose or other viscosity increasing solute. With the curvature concave to the T/η axis, the initial slope is related to the harmonic mean of the ρ values while the slope at high T/η in some cases approximates the arithmetic mean.

However, the ρ plots may produce curves which cannot readily be related to any of the above. This further contributes to the already complex task of interpreting nonlinear plots.

Several other problems may arise: (1) Some of the dye may be either adsorbed, free in solution, or "trapped" within the protein structure giving rise to erroneous polarizations. (2) As indicated in several of the studies already presented, the macroviscosity (solution viscosity) may be different from the microviscosity. Hence, in spite of obtaining either a linear or a curved ρ plot, the true plot may be either curved or linear, respectively.

2. *Excited State Lifetimes*

a. *Determination.* The main difficulty is the determination of the lifetimes for the specific compound under the specific conditions of interest. Few investigations of polarization of fluorescence have until now been immediately correlated with lifetime measurements (cf. Steiner and McAlister, 1957). It is possible that a part of the data collected and interpreted in the literature until 1969 has been based on incorrect lifetimes.

There are relatively simple methods whereby lifetimes can be obtained approximately, although certain assumptions are necessary. Weber (1952b) employed Oncley's value for the ρ for ovalbumin-DNS in order to determine the lifetime of DNS. Then, assuming the same value held for BSA-DNS, ρ of BSA was calculated and found to compare quite well with Oncley's value of BSA. This has become a standard procedure for determining the lifetime of a new dye; i.e., labeling BSA and assuming its ρ and determining τ from a ρ plot. The main difficulties with this approach are that ρ of the new BSA dye conjugate may be different (in spite of constancy of diffusion and sedimentation data), and that the lifetime of the dye on a different protein may be quite different. This can be partially checked by a comparison of the absorption and emission spectra and the quantum yield. If all three of these are the same, then there is a high probability that the lifetime is unchanged. However, if only the quantum yield is different, then several interpretations are possible to which we will give some consideration.

As pointed out by Weber (1953) it is too often assumed that there is a relationship between quantum yield and lifetime of the following nature:

$$q_1/q_2 = \tau_1/\tau_2 \tag{4.6}$$

This is true with collisional quenching (time independent) which competes with the fluorescence, and which is oftentimes the type of quenching which is taking place. However, the relationship need not be true if the quenching results from the formation of a dark complex. Weber (1948) has treated this situation with the resultant relationship emerging:

$$\frac{F_0}{F} = \frac{\tau_0}{\tau[1 + (t_c/\tau)(\tau_0/\tau - 1)]} \tag{4.7}$$

where F_0 = fluorescence intensity before quenching
F = fluorescence intensity after quenching
τ_0 = lifetime before quenching
τ = lifetime after quenching
t_c = mean life of the complex

It is seen that if t_c is much smaller than τ_0, then Eq. (4.6) results. As Weber (1948) suggested, deviations from Eq. (4.6) can be detected by plotting F/F_0 vs $1/p$. A straight line results if t_c/τ_0 is 10^{-2} or less. Hence, if the mean life of the complex is long compared to τ_0, then no change in lifetime need take place. Another type of quenching which need not change the lifetime is when the deactivation process can compete with the initial thermal relaxation process and not with the fluorescent step (Memming, 1961). In this case, the quantum yield can be reduced to any value without any change in the lifetime. In practice, Eq. (4.6) is useful in many circumstances. It should be noted, however, that the quantum yields must be determined correctly by integrating emission spectra from solutions of identical optical densities. With one dye this presents little difficulty if the lifetime of the dye in a particular solution is known. However, if a different dye is used as the standard to determine the absolute quantum yield of the dye in question, then the spectra must be corrected for the fluorometer efficiency at each wavelength (see Chapter 6 for this procedure). This factor has been neglected by some workers. From the above it should be clear that it is essential to determine the lifetime of each conjugate under the conditions of operation in order to properly interpret p plots under conditions which could change the lifetime such as (1) variable labeling, (2) denaturing agents, and (3) different proteins.

b. *Range of* τ. Weber (1952b) has shown that the following relationship exists:

$$(3+p_0)/3e > \rho/\tau_0 > (3+p_0)/(p_0/p_{min}-1) \qquad (4.8)$$

for excitation with natural light with p_{min} equal to the smallest measurable polarization with standard error e, which is $(p_0-p)/3p$. The analogous relationship for polarized excitation is

$$(3-p_0)/3e > \rho/\tau_0 > (3-p_0)/(p_0/p_{min}-1) \qquad (4.9)$$

It can be seen that, for a given lifetime, only a certain range of relaxation times can be determined. If the relaxation time is long compared to the lifetime, almost complete polarization results. If it is short, almost complete depolarization occurs. Hence, this constitutes a limitation on the proteins which can be studied for a particular dye. The longest accurately determined lifetime for a conjugated dye now known (1969) is that of DNS, which is around 12–14 nsec. There is a strong need for dyes with longer excited lifetimes for polarization studies of large macromolecules.

In summary, the studies with protein conjugates are limited by restricted combinations of lifetime, spectral characterics, and binding specificity. A review of fluorescent conjugates has been presented by Steiner and Edelhoch (1962). In Table 4.5 are summarized some characteristics of some of the dyes which are currently used in fluorescence polarization studies.

3. *Identity of the Fluorescent Molecule*

Most, if not all, fluorescent labeling agents are very slightly soluble in water; this requires the coupling reaction to be heterogeneous, i.e., to be executed by addition of the solid dye to the aqueous solution, by addition of a solution of the dye in a nonaqueous solvent to the aqueous medium, or by performing the reaction in an appropriate solvent mixture. All of these procedures can cause denaturation of the macromolecule. In addition, the dye molecule may cause changes in the structure through adsorption effects or by coupling with groups intimately associated with the structural integrity of the native protein (Weber and Young, 1964). In general, however, labeled proteins which are seemingly unchanged in biological activity, sedimentation, diffusion, and antibody specificity can be prepared. A possible situation which must be kept in mind is that where a protein is conjugated with approximately one dye per molecule of protein and with only 5% loss of activity, but where the 5% denatured protein may have taken on 15–20 dyes per molecule of protein, while the active protein is but

TABLE 4.5 RANGE OF DEPOLARIZATION METHOD[a]

Dye	Lifetime τ, nsec	Range of relaxation times ρ, nsec	Measurable molecular weight range $\times 10^{-3}$	Excitation wavelength	Emission wavelength
3-OH-pyrene 5:8:10: trisulfonic acid	90	4–250	40–2500	330	375
					420
DNS	12	0.5–30	5–300	340	500
3-Phenyl-7-iso-cyanato-coumarin	2.5	0.1–7	1–60	–	–
Anthracene iso-cyanate	15		100–300	350	400
				365	
				385	
Fluorescein isothio-cyanate	5	0.2–14	2–120	250	550
				285	
				325	
				495	
Acriflavine	4.4	0.176–12.3	1.8–110	380	
				470	
Rhodamine				250	605
				300	
				350	
				420	
				550	

[a]Adapted from Chadwick et al. (1960) and Steiner and Edelhoch (1962).

slightly labeled. Therefore, all measurements are on the inactive 5% of the protein. In the same vein, if 95% activity is maintained for a labeled preparation, one doesn't know what this signifies in terms of the original structure. Enzymatically it could mean that only very local alterations are present; conversely, significant changes could be present in portions of the molecule distant from the "active site." Once again, the need is demonstrated for independent measurements by other methods to ascertain changes in the native structure. Any significant changes, be they enzymatic, hydrodynamic, or spectral, should suggest a path of caution to be followed by the investigator. However, it is equally important to realize that constancy of such data does not necessarily prove that structural integrity has been maintained. As indicated by Gill (1965) fluorescence polarization can be a more sensitive measure of overall molecular rigidity than the other hydrodynamic methods. Hence, changes in the structure of the molecule can occur without being detected by those methods.

Several other limitations which will simply be noted, but which may be of importance are: (1) energy transfer between the pairs of protein-chromophores-dye and dye-dye, (2) concentration restrictions, (3) resolution of instruments, and (4) light scattering complications.

REFERENCES

Arkin, L., and Singleterry, C. (1949). *J. Colloid. Sci.*, **4**, 537.
Buchader, G., and Lebesque, J. (1946). *Compt. Rend.*, **223**, 324.
Chadwick, C. S., Johnson, P., and Richards, E. G. (1960). *Nature*, **186**, 239.
Churchich, J. (1963). *BBA*, **75**, 274.
Dandliker, W., Schapiro, H., Meduski, J., Alonso, R., Feigen, G., and Hamrick, J., (1965). *Immunochem.*, **1**, 165.
Gill, T. (1965). *Biopolymers*, **3**, 43.
Johnson, P., and Richards, E. (1962). *Arch. Biochem. Biophys.*, **97**, 260.
Kryszewsk, M., and Grosmonowa, B. (1961). *J. Polymer Sci.*, **52**, 85.
Laurence, D. (1952). *Biochem. J.*, **51**, 168.
Memming, R. (1961). *Z. Physik. Chem., Neue Folge*, **28**, 168.
Oncley, J. L. (1942). *Chem. Rev.*, **30**, 443.
Perrin, F. (1926). *J. Phys. Rad.*, **7**, 390.
Singleterry, C., and Weinberger, L. (1951). *JACS*, **73**, 4574.
Soleillet, P. (1929). *Ann Phys.* **12**, 23.
Steiner, R. F., and Edelhoch, H. (1962). *Chem. Revs.*, **62**, 457.
Steiner, R. F., and McAlister, A. (1957). *J. Colloid. Sci.*, **24**, 105.
Weber, G. (1948). *Trans. Faraday Soc.*, **44**, 185.
Weber, G. (1952a). *Biochem. J.*, **51**, 145.
Weber, G. (1952b). *Biochem. J.*, **51**, 155.
Weber, G. (1953). *Adv. Protein Chem.*, **8**, 415.
Weber, G., and Young, L. (1964). *J. Biol. Chem.*, **239**, 1424.
Weber, G., and Teale, F. (1965). *The Proteins*, **III**, 445.
Weber, G. (1966). In *Fluorescence and Phosphorescence Analyses* (D. M. Hercules, ed.), Wiley-Interscience, New York.

Appendix: Derivation of Polarization of Fluorescence Equations*

A4.1 Polarization—General Equations

A. EXCITING LIGHT POLARIZED

A general expression for the polarization of the light emitted from any single emission oscillator can be derived by reference to Fig. A4.1.

*The derivation of these equations arise primarily from the papers of G. Weber (1952a, b, 1953, 1956).

The total fluorescence intensity can be resolved into three components, I_x, I_y, and I_z, whose electric vectors are parallel to the three mutually perpendicular axes X, Y, and Z. With the maximum intensity normalized (equal to 1), each intensity component is equal to $\cos^2 \theta_i$ where θ_i is the angle between the oscillator and the component electric vector. (The $\cos^2 \theta_i$ relationship follows from the nature of dipole emission whereby the intensity is proportional to the square of the electric field strength, the latter of which is essentially proportional to $\cos \theta_i$ as θ_i is here defined). With the arrangement given in Fig. A4.1, the polarization is defined as

$$p = \frac{I_\| - I_\perp}{I_\| + I_\perp} = \frac{I_z - I_x}{I_z + I_x}$$

$I_\|$ is equal to $\cos^2 \theta$ and is independent of ω. $I_\perp$, however, is dependent upon both θ and ω. To determine the $I_\perp$ component, OE is projected

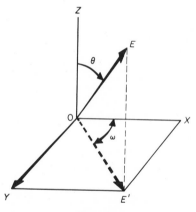

Fig. A4.1

onto the XY plane yielding OE′ which is equal to $\sin \theta$. OE′, in turn, is projected onto the X axis yielding $(\cos \omega)(\text{OE})$ or $(\cos \omega \sin \theta)$. Thus, $I_\perp$ is equal to $(\cos \omega \sin \theta)^2$. The polarization is equal to

$$p = \frac{\cos^2 \theta - \cos^2 \omega \sin^2 \theta}{\cos^2 \theta + \cos^2 \omega \sin^2 \theta} \qquad (A4.1)$$

If a collection of oscillators undergoes excitation, the observed polarization will be the above expression with the average value of the functions substituted:

$$p = \frac{\overline{\cos^2 \theta} - \overline{\cos^2 \omega} \; \overline{\sin^2 \theta}}{\overline{\cos^2 \theta} + \overline{\cos^2 \omega} \; \overline{\sin^2 \theta}} \qquad (A4.2)$$

With polarized excitation and a random collection of absorption oscillators, however, the Z axis is an axis of symmetry whereby all values of ω from 0 through 360° are equally probable (see Section A4.2A). Therefore, $\overline{\cos^2 \omega}$ is calculated to be:

$$\overline{\cos^2 \omega} = \frac{\int_0^{2\pi} \cos^2 \omega \, d\omega}{\int_0^{2\pi} d\omega} = \frac{1}{2} \tag{A4.3}$$

and

$$p = \frac{\overline{\cos^2 \theta} - \frac{1}{2}\overline{\sin^2 \theta}}{\overline{\cos^2 \theta} + \frac{1}{2}\overline{\sin^2 \theta}} \tag{A4.4}$$

By the following rearrangements and remembering that $\overline{\sin^2 \theta} + \overline{\cos^2 \theta} = 1$, the final equation (A4.7), which will be used for subsequent derivations, results:

$$p = \frac{\overline{\cos^2 \theta} - \frac{1}{2}(1 - \overline{\cos^2 \theta})}{\overline{\cos^2 \theta} + \frac{1}{2}(1 - \overline{\cos^2 \theta})} = \frac{\frac{3}{2}(\overline{\cos^2 \theta} - \frac{1}{2})}{\frac{1}{2}(\overline{\cos^2 \theta} + \frac{1}{2})} = \frac{3\overline{\cos^2 \theta} - 1}{\overline{\cos^2 \theta} + 1} \tag{A4.5}$$

$$\frac{1}{p} - \frac{1}{3} = \frac{\overline{\cos^2 \theta} + 1}{3\overline{\cos^2 \theta} - 1} - \frac{1}{3} = \frac{3\overline{\cos^2 \theta} + 3 - 3\overline{\cos^2 \theta} + 1}{3(3\overline{\cos^2 \theta} - 1)} \tag{A4.6}$$

$$\frac{1}{p} - \frac{1}{3} = \frac{\frac{2}{3}}{(3\overline{\cos^2 \theta} - 1)/2} \tag{A4.7}$$

For any collection of emission oscillators it is therefore only necessary to determine $\overline{\cos^2 \theta}$ in order to calculate or derive a further expression for p.

Two limits of Eq. (A4.5) can be readily verified. For a random collection of emission oscillators, the polarization must be zero. $\overline{\cos^2 \theta}$ for this case, is calculated below:

$$\overline{\cos^2 \theta} = \frac{\int_0^{\pi/2} \cos^2 \theta \sin \theta \, d\theta}{\int_0^{\pi/2} \sin \theta \, d\theta} = \frac{1}{3} \tag{A4.8}$$

where the number of oscillators at angle θ is proportional to $\sin \theta$ (see Section A4.2A) and θ assumes all values from 0 through $\pi/2$.* Upon substituting $\frac{1}{3}$ into Eq. (A4.5).

$$p = \frac{3(\frac{1}{3}) - 1}{\frac{1}{3} + 1} = 0$$

---

*As oscillators have no head or tail, $0 = 0'$ is equivalent to $0 = 180°$. Therefore, all orientations are covered by $0 = 0 \rightarrow 90°$ and $\omega = 0 \rightarrow 360°$.

In the other limit, if all the oscillators are parallel to the Z axis, for example, the polarization must be 1. This is easily seen when the value of $\overline{\cos^2\theta} = 1$ is used in Eq. (A4.5). This yields:

$$p = \frac{3(1)-1}{1+1} = 1$$

Another valuable expression can be derived from Eq. (A4.7). If there are n different species of emission oscillators, each specie with its own $\overline{\cos^2\theta_i}$ the final observed value, $\overline{\cos^2\theta}_{\text{final}}$ is

$$\overline{\cos^2\theta}_{\text{final}} = \sum_{i=1}^{i=n} \overline{\cos^2\theta_i} \cdot f_i \tag{A4.9}$$

where f_i is the fraction of the total fluorescence intensity contributed by the ith specie. Therefore, by substituting Eq. (A4.9) into Eq. (A4.7)

$$\frac{1}{p}-\frac{1}{3} = \frac{\frac{2}{3}}{\frac{3}{2}(\Sigma\,\overline{\cos^2\theta_i}\cdot f_i)-\frac{3}{2}} \tag{A4.10}$$

which can be rearranged as follows

$$\frac{\frac{2}{3}}{\frac{3}{2}(\Sigma\,\overline{\cos^2\theta_i}\cdot f_i)-\frac{3}{2}} = \frac{\frac{2}{3}}{\frac{1}{2}\Sigma(3\,\overline{\cos^2\theta_i}\cdot f_i)-\frac{1}{2}\Sigma f_i}$$

because $3\Sigma X_i = \Sigma 3X_i$ and $\Sigma f_i = 1$. Factoring out f_i and further rearranging yields Eq. (A4.11):

$$\frac{1}{p}-\frac{1}{3} = \sum \frac{\frac{2}{3}}{\frac{1}{2}f_i(3\,\overline{\cos^2\theta_i}-1)} \quad \sum \frac{1/f_i}{\frac{3}{2}\,\overline{\cos^2\theta_i}-\frac{1}{2}} \tag{A4.11}$$

But

$$\frac{1}{p_i}-\frac{1}{3} = \frac{\frac{2}{3}}{\frac{3}{2}\,\overline{\cos^2\theta_i}-\frac{1}{2}} \tag{A4.12}$$

Therefore,

$$\frac{1}{p}-\frac{1}{3} = \left(\sum \frac{f_i}{\frac{1}{p_i}-\frac{1}{3}}\right)^{-1} \tag{A4.13}$$

Defining anisotropy A as follows

$$A = \left(\frac{1}{p}-\frac{1}{3}\right)^{-1} \tag{A4.14}$$

Eq. (4.13) can be written as

$$A = \sum f_i A_i \qquad (A4.15)$$

B. EXCITING LIGHT UNPOLARIZED

An analogous set of equations may be derived when the excitation is with unpolarized light. As the electric vectors of the exciting light are all in the YZ plane, the X axis becomes an axis of symmetry. Therefore, the angles are designated differently as seen in Fig. A4.2. In geometric terms the intensities of $I_\parallel$ and $I_\perp$ have the following values:

$$I_\parallel = \sin^2\theta \sin^2\omega$$
$$I_\perp = \cos^2\theta$$

where $I_\parallel$ and $I_\perp$ are again the Z- and X-axis components of the fluorescent light.

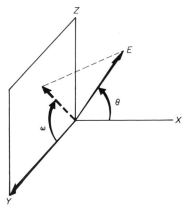

Fig. A.4.2

Therefore,

$$p = \frac{\sin^2\theta \sin^2\omega - \cos^2\theta}{\sin^2\theta \sin^2\omega + \cos^2\theta} \qquad (A4.16)$$

With a large number of oscillators, an expression analogous to (A4.2) is obtained:

$$p = \frac{\overline{\sin^2\theta}\,\overline{\sin^2\omega} - \overline{\cos^2\theta}}{\overline{\sin^2\theta}\,\overline{\sin^2\omega} + \overline{\cos^2\theta}} \qquad (A4.17)$$

However, $\overline{\sin^2\omega} = \tfrac{1}{2}$ as previously shown with $\overline{\cos^2\omega}$ (A4.3).

Therefore,

$$p = \frac{\frac{1}{2}\overline{\sin^2\theta} - \overline{\cos^2\theta}}{\frac{1}{2}\overline{\sin^2\theta} + \overline{\cos^2\theta}} \qquad (A4.18)$$

This can be rearranged into a more useful form to give an equation analogous to Eq. (A4.7). Inverting, adding $\frac{1}{3}$, collecting terms, and multiplying top and bottom by -1 gives the following.

$$\frac{1}{p} + \frac{1}{3} = \frac{(1 - \overline{\cos^2\theta}) + 2\overline{\cos^2\theta}}{(1 - \overline{\cos^2\theta}) - 2\overline{\cos^2\theta}} + \frac{1}{3} = \frac{1 + \overline{\cos^2\theta}}{1 - 3\overline{\cos^2\theta}} + \frac{1}{3}$$

$$= \frac{4}{3(1 - 3\overline{\cos^2\theta})} = \frac{-4}{3(\overline{\cos^2\theta} - 1)}$$

Dividing top and bottom by 2 and rearranging yields the desired expression (A4.19):

$$\frac{1}{p} + \frac{1}{3} = \frac{-\frac{2}{3}}{\frac{3}{2}\overline{\cos^2\theta} - \frac{1}{2}} \qquad (A4.19)$$

Also, as before,

$$\cos^2\theta_{\text{final}} = \sum_{i=1}^{i=n} f_i \cdot \overline{\cos^2\theta_i}$$

Therefore,

$$\frac{1}{p} + \frac{1}{3} = \frac{-\frac{2}{3}}{\frac{3}{2}(\Sigma f_i \cdot \overline{\cos^2\theta_i}) - \frac{1}{2}} = \left(\sum \frac{f_i/(-\frac{2}{3})}{\frac{3}{2}\overline{\cos^2\theta_i} - \frac{1}{2}}\right)^{-1} = \left(\sum \frac{f_i}{(1/p_i) + (\frac{1}{3})}\right)^{-1} \qquad (A4.20)$$

Defining A as

$$A = \left(\frac{1}{p} + \frac{1}{3}\right)^{-1} \qquad (A4.21)$$

results in

$$A = \sum f_i A_i \qquad (A4.22)$$

which is identical to the expression for polarized excitation (A4.15).

A4.2 Absorption Anisotropy

A. Exciting Light Polarized

Two factors determine the average orientation of the excited absorption oscillators. The first (Fig. A4.1) is that the probability of excitation is proportional to $\cos^2\theta$. The second is that the number of dipoles at any given θ increases with θ. This is easily seen by reference to Fig. (A4.3). A random distribution of the dipoles will describe a series of concentric cones, each cone representing those oscillators with given

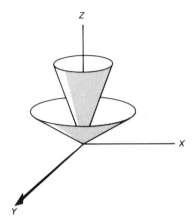

Fig. A4.3

values of θ. As the density of oscillators must be equal on all cone surfaces, the number of oscillators at each θ is proportional to the surface area (s) of the cone where $S = \pi h r = \pi h \sin\theta$. Therefore, the first factor (probably of absorption) gives greater weight to the smaller angles, but the second factor (number of oscillators) favors the larger angles. The relative number of oscillators (No.) excited at each θ must be the product of those two factors, or the probability of excitation [Prob (θ)] times the number of oscillators at that angle [No.(θ)]:

$$\text{No.} = \text{No.}(\theta) \times \text{Prob.}\,(\theta) \tag{A4.23}$$

To average $\cos^2\theta$ one must integrate $\cos^2\theta$ times its weighting factors and divided by the integrated weighting factors, as given in (A4.24). Therefore,

$$\overline{\cos^2\theta} = \frac{\int_0^{\pi/2} \cos^2\theta \times \text{Prob}\,(\theta) \times \text{No.}(\theta)}{\int_0^{\pi/2} \text{Prob}\,(\theta) \times \text{No.}(\theta)} \tag{A4.24}$$

Or,

$$\overline{\cos^2\theta} = \frac{\int_0^{\pi/2} \cos^2\vartheta \times \cos^2\theta \times \sin\theta \, d\theta}{\int_0^{\pi/2} \cos^2\theta \times \sin\theta \, d\theta} = \frac{3}{5}$$

If the distribution of excited oscillators were random, $\overline{\cos^2\theta}$ would equal $\frac{1}{2}$ as seen in Eq. (A4.3). However, the value of $\overline{\cos^2\theta} = \frac{3}{5}$ indicates the favored orientation toward the Z axis. Another fact is apparent from Fig. (A4.3). The Z axis is an axis of symmetry, such that the excited state distribution is identical from any point of observation in the XY plane. Hence, polarization measurements may be made at any angle to the X axis, not necessarily at 90°, to obtain equal results. In addition, it is apparent now how the initial polarization of the exciting light is changed by the absorption process.

B. EXCITING LIGHT UNPOLARIZED

At first it may be difficult to see how excitation with completely unpolarized light could lead to a nonrandom distribution of excited oscillators. However, the random nature of the vector distribution is only in the YZ plane. As viewed from the Y axis, if this were possible without reflection or an absorption-emission process, there is no electric vector component in the XZ plane. Hence, it can be said that the radiation is completely polarized from the perspective of right-angle observation. Thus, the comparison with the polarized excitation is more easily discerned. Referring to Fig. A4.2, it is seen that when the electric vectors of the exciting light are all in the YZ plane, those oscillators that are in that plane are preferentially excited, and oscillators with equal θ's have equal probability to be excited. In addition, with reference to Fig. A4.4, the number of oscillators at a given angle θ again increases with θ. Hence both the number of dipoles and the probability of excitation increase as θ increases. It is now evident how the process of absorption results in a nonrandom distribution of excited oscillators. Also, the concentric cones of these oscillators are coaxial to the X axis, making this an axis of symmetry. Hence,

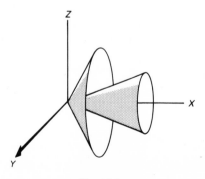

Fig. A4.4

observations made from any point in the *YZ* plane will give identical results. Analogously to polarized excitation, the following relationships hold:

Probability of absorption $\propto \sin^2\theta$
Number of oscillators $\propto \sin \theta$

Therefore,

$$\overline{\cos^2\theta} = \frac{\int_0^{\pi/2} \cos^2\theta \sin^2\theta \sin \theta \, d\theta}{\int_0^{\pi/2} \sin^2\theta \sin \theta \, d\theta} \tag{A4.25}$$

Upon integration, this yields

$$\overline{\cos^2\theta} = 1/5 \tag{A4.26}$$

It should be emphasized that θ is defined differently for polarized and nonpolarized excitation.

A4.3 Polarization of Immobile Emission Oscillators

A. PARALLEL EXCITATION AND EMISSION OSCILLATORS

1. *Polarized Excitation*

When the absorption and emission oscillators are parallel and no rotation of the emission oscillator occurs, Fig. A4.1 suffices for the description with OE representing the orientation of both oscillators. It is necessary only to determine $\overline{\cos^2\theta}$ of the emission oscillators and then use Eq. (A4.7). However, as the absorption and emission oscillators are parallel, $\overline{\cos^2\theta}$ (absorption oscillators) equals $\overline{\cos^2\theta}$ (emission oscillators). It was determined in the last section that $\overline{\cos^2\theta}$ (absorption oscillators) $= \frac{3}{5}$. Substituting this value into Eq. A4.7 yields:

$$\frac{1}{p_0} - \frac{1}{3} = \frac{\frac{2}{3}}{[3(\frac{3}{5}) - 1]/2} = \frac{20}{9(3) - 15} = \frac{20}{12} = \frac{5}{3} \tag{A4.27}$$

and

$$p_0 = \tfrac{1}{2}$$

where p_0 is the polarization in the absence of any rotation of the emission oscillator.

2. Unpolarized

In the case of unpolarized excitation, one substitutes from Eq. (A4.26) $\overline{\cos^2\theta}$ (absorption) $= \frac{1}{5} = \overline{\cos^2\theta}$ (emission) and obtains

$$\frac{1}{p_0} + \frac{1}{3} = \frac{-\frac{2}{3}}{[3(\frac{1}{5}) - 1]/2} = \frac{10}{3} \qquad (A4.28)$$

and

$$p_0 = \frac{1}{3}$$

Therefore, if a random distribution of absorption oscillators are excited, these polarizations, $p_0 = \frac{1}{2}$ and $p_0 = \frac{1}{3}$, represent the maximum values of polarization for polarized and natural light excitation, respectively. Other possible cases lead to different polarizations. Two such cases are (1) nonparallel oscillators and (2) movements of the excited emission oscillators. In the next section the polarization equations for these cases will be derived.

B. NONPARALLEL ABSORPTION AND EMISSION OSCILLATORS

1. Soleillet's Equation

Figure A4.5 represents the situation when the oscillators are not parallel, where α represents the angle between the electric vector of the exciting light (OZ) and the absorption (OA) oscillator while λ represents the angle between the absorption (OA) and emission (OE) oscillators. For any given values of θ and ω (Fig. A4.1) OE can be located with equal probability at any point along EFE (Fig. A4.5) due to the random distribution of molecules. It is obvious that for each

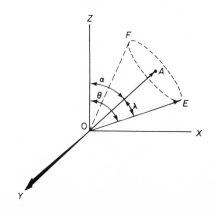

Fig. A4.5

value of θ there is not a unique value for θ and, subsequently, for $\cos^2\theta$. An easy solution to this problem, however, is offered by Soleillet's equation (Soleillet, 1929) and is derived as follows.

With reference to Fig. A4.5 assume OZ undergoes a displacement α to OA and then OA another displacement λ to OE with planes ZOA and AOE not required to be coplanar. The angle between OZ and OE, θ, is given by the following trigonometric function:

$$\cos\theta = \cos\alpha\cos\lambda + \sin\alpha\sin\lambda\cos\gamma \qquad (A4.29)$$

where γ is the angle between the two planes ZOA and AOE. If these angular displacements occur a large number of times, squaring and taking averages yields

$$\overline{\cos^2\theta} = \overline{\cos^2\alpha}\;\overline{\cos^2\lambda} + 2\;\overline{\cos\alpha}\;\overline{\cos\lambda}\;\overline{\sin\alpha}\;\overline{\sin\lambda}\;\overline{\cos\gamma}$$
$$+ \overline{\sin^2\alpha}\;\overline{\sin^2\lambda}\;\overline{\cos^2\gamma} \qquad (A4.30)$$

But as all values of γ from 0 to 2π are equally probable, i.e., the transition from OA to OE is isotropic,

$$\overline{\cos\gamma} = \frac{\int_0^{2\pi}\cos\gamma\,d\gamma}{\int_0^{2\pi}d\gamma} = 0 \qquad (A4.31)$$

and

$$\overline{\cos^2\gamma} = \frac{\int_0^{2\pi}\cos^2\gamma\,d\gamma}{\int_0^{2\pi}d\gamma} = \frac{1}{2} \qquad (A4.32)$$

Therefore, for a large number of displacements, the average displacement is defined by the following:

$$\overline{\cos^2\theta} = \overline{\cos^2\alpha}\;\overline{\cos^2\lambda} + \tfrac{1}{2}\overline{\sin^2\alpha}\;\overline{\sin^2\lambda} \qquad (A4.33)$$

This can be rearranged to a more useful form as follows:

$$\overline{\cos^2\theta} = \overline{\cos^2\alpha}\;\overline{\cos^2\lambda} + \tfrac{1}{2}(1 - \overline{\cos^2\alpha})(1 - \overline{\cos^2\lambda}) \qquad (A4.34)$$

Multiplying by 6 and subtracting 2 from each side yields:

$$2(3\,\overline{\cos^2\theta} - 1) = (3\,\overline{\cos^2\alpha} - 1)(3\,\overline{\cos^2\lambda} - 1) \qquad (A4.35)$$

Dividing by 4, the final result is

$$\left(\frac{3\,\overline{\cos^2\theta}-1}{2}\right)=\left(\frac{3\,\overline{\cos^2\alpha}-1}{2}\right)\left(\frac{3\,\overline{\cos^2\lambda}-1}{2}\right) \qquad \text{(A4.36)}$$

Hence, the addition of two displacements results in the multiplication of the quantities:

$$\left(\frac{3\,\overline{\cos^2\lambda}-1}{2}\right)\left(\frac{3\,\overline{\cos^2\alpha}-1}{2}\right)$$

It can be shown that if another isotropic displacement occurred from OE to OG, through an angle δ, this would merely add another term such that:

$$\left(\frac{3\,\overline{\cos^2\theta}-1}{2}\right)=\left(\frac{3\,\overline{\cos^2\alpha}-1}{2}\right)\left(\frac{3\,\overline{\cos^2\lambda}-1}{2}\right)\left(\frac{3\,\overline{\cos^2\delta}-1}{2}\right) \quad \text{(A4.37)}$$

with the only stipulation being that each displacement after the first occur isotropically; i.e., with equal probability in all directions.

2. Polarized Excitation

It will now be seen how Eq. (A4.36) can be utilized in Eq. (A4.7) to give

$$\frac{1}{p_0}-\frac{1}{3}=\frac{\frac{2}{3}}{\frac{1}{4}(3\,\overline{\cos^2\alpha}-1)(3\,\overline{\cos^2\lambda}-1)} \qquad \text{(A4.38)}$$

where p_0 refers to the polarization in the absence of rotation of the emission oscillator.

It has already been shown that in the case of polarized excitation (Section A4.2)

$$\overline{\cos^2\alpha}=\tfrac{3}{5}$$

Therefore,

$$\frac{1}{p_0}-\frac{1}{3}=\frac{\frac{2}{3}}{\frac{1}{4}(3(\frac{3}{5})-1)(3\,\overline{\cos^2\lambda}-1)}=\frac{5}{3}\frac{2}{3\,\cos^2\lambda-1} \qquad \text{(A4.39)}$$

Rearrangement of this equation gives

$$p_0=\frac{3\,\cos^2\lambda-1}{\cos^2\lambda+3} \qquad \text{(A4.40)}$$

where λ is the angle between the absorption and emission oscillators.

When $\lambda = 0$, $p = \frac{1}{2}$, which is the case for parallel absorption and emission oscillators. When $\lambda = \pi/2, p = -\frac{1}{3}$. Therefore,

$$\tfrac{1}{2} \geqslant p_0 \geqslant -\tfrac{1}{3} \qquad (A4.41)$$

3. *Unpolarized Excitation*

In Fig. A4.6 the analogous situation for unpolarized excitation is described. Therefore,

$$\frac{1}{p_0} + \frac{1}{3} = \frac{-\frac{2}{3}}{\frac{1}{4}(3\,\overline{\cos^2\alpha})(3\,\overline{\cos^2\lambda} - 1)} \qquad (A4.42)$$

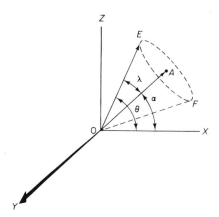

Fig. A4.6

But from Eq. (A4.26), $\overline{\cos^2\theta} = \frac{1}{5}$, and

$$\frac{1}{p_0} + \frac{1}{3} = \frac{-\frac{2}{3}}{\frac{1}{4}(3(\frac{1}{5}) - 1)(3\,\overline{\cos^2\lambda} - 1)} = \frac{20}{3}\left(\frac{1}{3\,\overline{\cos^2\lambda} - 1}\right) \qquad (A4.43)$$

or

$$p_0 = \frac{3\,\overline{\cos^2\lambda} - 1}{7 - \overline{\cos^2\lambda}} \qquad (A4.44)$$

When $\lambda = 0°, p_0 = \frac{1}{3}$, and when $\lambda = \pi/2, p_0 = -\frac{1}{7}$. Hence,

$$\tfrac{1}{3} \geqslant p_0 \geqslant -\tfrac{1}{7} \qquad (A4.45)$$

A4.4 Polarization of Mobile Emission Oscillators

A. BROWNIAN MOTION

If each emission oscillator OE rotates over a certain angle β (Fig. A4.7) before emission, the polarization will clearly change. For the derivation of the equations describing this effect, however, it is necessary to add only the term $(\frac{3}{2}\overline{\cos^2\beta} - 1)$ to Eq. (A4.38) (polarized excitation), yielding

$$\frac{1}{p} - \frac{1}{3} = \frac{\frac{2}{3}}{\frac{1}{6}(3\overline{\cos^2\alpha} - 1)(3\overline{\cos^2\lambda} - 1)(3\overline{\cos^2\beta} - 1)}$$

$$= \left(\frac{1}{p_0} - \frac{1}{3}\right)\left(\frac{2}{3\overline{\cos^2\beta} - 1}\right) \qquad (A4.46)$$

The overall rotation, and therefore $\overline{\cos^2\beta}$, is dependent upon two factors: (1) how far a given oscillator rotates in a certain time interval, and (2) how many oscillators emit at a certain time t after excitation.

The rotation of a molecule undergoing Brownian motion was treated by Einstein (1906) with the following result:

$$\overline{\beta_{\delta t}^2} = \frac{4\bar{E}\,\delta t}{f_r} \qquad (A4.47)$$

where δt is a small unit of time, $\overline{\beta_{\delta t}^2}$ is the square of the average angular displacement during δt, $\bar{E}$ is the average kinetic energy possessed by the molecule over this unit of time, and f_r is the frictional coefficient of the rotating species (to which, for our purposes, the oscillator is rigidly attached). As δt is small, $\overline{\beta_{\delta t}^2}$ is small; therefore, $\overline{\sin^2\beta_{\delta t}} = \overline{\beta_{\delta t}^2}$ and

$$1 - \overline{\cos^2\beta_{\delta t}} = 3\bar{E}\,\delta t/f_r \qquad (A4.48)$$

As the rotation of a dipole occurs around two axes, $\bar{E} = kT$ with

$$\overline{\cos^2\beta_{\delta t}} = 1 - 4kT/f_r \qquad (A4.49)$$

which is rearranged to

$$\frac{3\overline{\cos^2\beta_{\delta t}} - 1}{2} = 1 - 6kT\delta t/f_r \qquad (A4.50)$$

Visualizing each δt as giving rise to a discrete displacement, $\overline{\cos^2\beta_{\delta t}}$, the rotation over a period of time t, described as $\overline{\cos^2\beta_t}$, where there are $t/\delta t$ discrete rotations, is described by Soleillet's equation as:

$$\left(\frac{3\,\overline{\cos^2\beta_t}-1}{2}\right) = \left(\frac{3\,\cos^2\beta_{\delta t}-1}{2}\right)^{t/\delta t} \tag{A4.51}$$

By substitution, one obtains

$$\left(\frac{3\,\overline{\cos^2\beta_t}-1}{2}\right) = \left(1-\frac{6\,kT\delta t}{f_r}\right)^{t/\delta t} \tag{A4.52}$$

With $\delta t/t \ll 1$, expansion of the exponential leads to

$$\frac{3\,\overline{\cos^2\beta_t}-1}{2} = e^{-6kTt/f_r} \tag{A4.53}$$

or

$$\overline{\cos^2\beta_t} = \tfrac{1}{3} + \tfrac{2}{3}\,e^{-6kTt/f_r} \tag{A4.54}$$

This rotation may be described in terms of a constant ρ, which is given the name "rotational relaxation time." This constant is dependent upon both the solvent properties and the molecules or rotating species, itself, and has the dimensions of time:

$$\rho = \frac{f_r}{2kT} = \frac{8\pi\eta r^3}{2kT} \tag{A4.55}$$

where η is the viscosity, r is the radius of the equivalent sphere, T is the temperature, k is the Boltzmann constant, and f_r is the frictional coefficient. Therefore, ρ is a constant for a given molecule only under certain conditions governing its structure and determining T/η for the solution. With the above definition, Eq. (A4.54) can be rewritten as follows:

$$\overline{\cos^2\beta_t} = \tfrac{1}{3} + \tfrac{2}{3}e^{-3t/\rho} \tag{A4.56}$$

If all molecules are emitting at the same time t, this relationship could be used directly to describe the polarization. However, since fluorescence is an exponential decay phenomenon, molecules will be fluorescing at different times, and, consequently, after different degrees of rotation. Therefore, the final average rotation will be the average of the rotations occurring over different intervals of time t, weighted according to the contribution $f(t)$ each rotation makes to the total fluorescence intensity. Hence

$$\overline{\cos^2\beta} = \int_0^\infty f(t)\,\overline{\cos^2\beta_t}\,dt \tag{A4.57}$$

The intensity I_t at time t for exponential decay (see Section 5.8) is described by

$$I_t = I_0 e^{-t/\tau} \tag{A4.58}$$

where I_0 is the intensity of the fluorescence at $t = 0$, and τ is the fluorescent lifetime or the average time spent in the excited state as described by the following expression:

$$\tau = \frac{\int_0^\infty t I_t \, dt}{\int_0^\infty I_t \, dt} \tag{A4.59}$$

The total fluorescence, however, is

$$I_T = \int_0^\infty I_0 e^{-t/\tau} \, dt \tag{A4.60}$$

Therefore, the fraction of the total fluorescence being emitted at time $t, f(t)$, is:

$$f(t) = \frac{I_t}{I_T} = \frac{I_0 e^{-t/\tau}}{\int_0^\infty I_0 e^{-t/\tau} \, dt} = \frac{e^{-t/\tau}}{\tau} \tag{A4.61}$$

Substitution of Eq. (A4.56) and (A4.61) into (A4.57) yields:

$$\overline{\cos^2 \beta} = \int_0^\infty \left(\frac{e^{-t/\tau}}{\tau} \right) \left(\tfrac{1}{3} + \tfrac{2}{3} e^{-3t/\rho} \right) dt \tag{A4.62}$$

or

$$\overline{\cos^2 \beta} = \frac{1}{3\tau} \int_0^\infty \left(e^{-t/\tau} + 2 e^{-t(3/\rho + 1/\tau)} \right) dt \tag{A4.63}$$

Upon integration,

$$\overline{\cos^2 \beta} = \frac{1}{3} + \frac{2}{3\tau} \left(\frac{1}{3/\rho + 1/\tau} \right) = \frac{1}{3} + \frac{2}{3} \left(\frac{\rho}{3\tau + \rho} \right) \tag{A4.64}$$

By substituting this expression into Eq. (A4.46) for $\overline{\cos^2 \beta}$ and simplifying, the final equation is obtained:

$$\frac{1}{p} - \frac{1}{3} = \left(\frac{1}{p_0} - \frac{1}{3} \right) \left(\frac{2}{3(\tfrac{1}{3} + \tfrac{2}{3}[\rho/(3\tau + \rho)]) - 1} \right) = \left(\frac{1}{p_0} - \frac{1}{3} \right) \left(1 + \frac{3\tau}{\rho} \right) \tag{A4.65}$$

Another useful equation is obtained by substituting the following expression as given in Eq. (A4.55) for ρ:

$$\rho = \frac{8\pi \eta r^3}{2kT} \tag{A4.55}$$

to obtain

$$\left(\frac{1}{p}-\frac{1}{3}\right) = \left(\frac{1}{p_0}-\frac{1}{3}\right)\left(1+\frac{3\tau kT}{4\pi\eta r^3}\right) \qquad \text{(A4.66)}$$

But the molecular volume of a sphere, V_0, is given by the following:

$$V_0 = \tfrac{4}{3}\pi r^3 \qquad \text{(A4.67)}$$

Hence

$$\left(\frac{1}{p}-\frac{1}{3}\right) = \left(\frac{1}{p_0}-\frac{1}{3}\right)\left(1+\frac{kT\tau}{V_0\eta}\right) \qquad \text{(A4.68)}$$

In terms of the molar volume, V_m:

$$\left(\frac{1}{p}-\frac{1}{3}\right) = \left(\frac{1}{p_0}-\frac{1}{3}\right)\left(1+\frac{RT\tau}{V_m\eta}\right) \qquad \text{(A4.69)}$$

This equation (A4.69) was first derived by Perrin (1926) and is usually referred to as Perrin's equation.

The derivation for excitation with natural or unpolarized light is exactly analogous and yields the following equations:

$$\left(\frac{1}{p}+\frac{1}{3}\right) = \left(\frac{1}{p_0}+\frac{1}{3}\right)\left(1+\frac{3\tau}{\rho}\right) \qquad \text{(A4.70)}$$

and

$$\left(\frac{1}{p}+\frac{1}{3}\right) = \left(\frac{1}{p_0}+\frac{1}{3}\right)\left(1+\frac{RT\tau}{V_m\eta}\right) \qquad \text{(A4.71)}$$

B. ENERGY TRANSFER

When the excitation energy possessed by one oscillator is transferred to another oscillator, further depolarization occurs (see Chapter 6). It is easiest to develop the relationships governing this phenomenon when negligible depolarization occurs through Brownian rotation. In that case, the change in the angle of the emission dipole is due only to energy transfer and can be described as angle β_{Et} (equal to β) in Fig. A4.7. Hence, it is only necessary to determine $\overline{\cos^2\beta_{Et}}$. With each individual transfer being independent and isotropic and giving an average change of $\overline{\cos^2\beta_A}$, the total change in angular orientation due to this transfer for an individual quantum is the cumulative effect of the total number of transfers, as described by Soleillet's equation. The final angle between the initially excited emission oscillator and the final emitting oscillator is equal to the following:

$$\frac{3\,\overline{\cos^2\beta_{Et(n)}}-1}{2} = \left(\frac{3\,\overline{\cos^2\beta_A}-1}{2}\right)^n \qquad \text{(A4.72)}$$

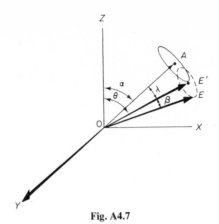

Fig. A4.7

where $\overline{\cos^2\beta_{Et(n)}}$ refers to the displacement after n transfers, $(\overline{\cos^2\beta_{Et}}$ would equal $\overline{\cos^2\beta_{Et(n)}})$. However, as was seen with Brownian motion, it is necessary to weigh the different angular displacements (different values of n) according to the contribution, $f(n)$ each one makes to the total intensity. Therefore,

$$\overline{\cos^2\beta_{Et}} = \sum_{n=0}^{n=\infty} f(n)\,\overline{\cos^2\beta_{Et(n)}} = \sum f(n)\left[\frac{2}{3}\left(\frac{3\,\overline{\cos^2\beta_A}-1}{2}\right)^n + \frac{1}{3}\right] \quad (A4.73)$$

$f(n)$, however, is equivalent to the probability of emission after n transfers as this probability is equivalent to the fraction of the dipoles emitting after n transfers or the fractional contribution to the total intensity. Assuming that all transfers occur independently with a probability of $\bar{q}$, (as opposed to the probability of emission, $[1-\bar{q}]$) then the probability of n transfers followed by emission is

$$f(n) = \frac{n}{\bar{q}}(1-\bar{q}) \quad (A4.74)$$

Hence,

$$\overline{\cos^2\beta_{Et}} = \sum_{n=0}^{n=\infty} \bar{q}^n(1-\bar{q})\left[\frac{2}{3}\left(\frac{3\,\overline{\cos^2\beta_A}-1}{2}\right)^n + \frac{1}{3}\right] \quad (A4.75)$$

which is rearranged to give terms described by $\sum ar^n$:

$$\overline{\cos^2\beta_{Et}} = \sum \tfrac{2}{3}(1-\bar{q})\left[\bar{q}\left(\frac{3\,\overline{\cos^2\beta_A}-1}{2}\right)\right]^n + \sum \tfrac{1}{3}(1-\bar{q})\bar{q}^n \quad (A4.76)$$

The determination of such a series when $r < 1$ is described by the following equation:

$$\sum ar^n = a + ar^1 + ar^2 + ar^3 + \cdots = \frac{a}{1-r} \qquad (A4.77)$$

Therefore,

$$\overline{\cos^2\beta_{Et}} = \frac{\frac{2}{3}(1-\bar{q})}{1-\bar{q}\left(\dfrac{3\,\overline{\cos^2\beta_A}-1}{2}\right)} + \frac{1}{3}\frac{(1-\bar{q})}{(1-\bar{q})} \qquad (A4.78)$$

or

$$\overline{\cos^2\beta_{Et}} = \frac{\frac{4}{3}(1-\bar{q})}{2-\bar{q}(3\,\overline{\cos^2\beta_A}-1)} + \frac{1}{3} \qquad (A4.79)$$

Therefore,

$$\frac{1}{p} - \frac{1}{3} = \left(\frac{1}{p_0} - \frac{1}{3}\right)\left[3\left(\frac{\frac{4}{3}(1-\bar{q})}{2-\bar{q}(3\,\overline{\cos^2\beta_A}-1)} + \frac{1}{3}\right) - 1 \right]$$

$$= \left(\frac{1}{p_0} - \frac{1}{3}\right)\left[\frac{4-6\bar{q}\,\overline{\cos^2\beta_A}+2\bar{q}}{4(1-\bar{q})}\right] \qquad (A4.80)$$

Dividing by 4 and substituting $(1-\overline{\sin^2\beta_A})$ for $\overline{\cos^2\beta_A}$ yields:

$$\frac{1}{p} - \frac{1}{3} = \left(\frac{1}{p_0} - \frac{1}{3}\right)\left[\frac{1-\frac{3}{2}\bar{q}+\frac{3}{2}\bar{q}\,\overline{\sin^2\theta_A}+\bar{q}/2}{1-\bar{q}}\right] \qquad (A4.81)$$

Finally

$$\frac{1}{p} - \frac{1}{3} = \left[\frac{1}{p_0} - \frac{1}{3}\right]\left[1 + \frac{3}{2}\overline{\sin^2\theta_A}\frac{\bar{q}}{1-\bar{q}}\right] \qquad (A4.82)$$

However, as each quantum is emitted only once, the average number of transfers is the ratio of the probability of transfer and the probability of emission, or

$$\bar{\eta} = \frac{\bar{q}}{1-\bar{q}} = \frac{\text{relative number of transfers}}{\text{relative number of emissions}} \qquad (A4.83)$$

Hence,

$$\frac{1}{p} - \frac{1}{3} = \left(\frac{1}{p_0} - \frac{1}{3}\right)(1 + \frac{3}{2}\overline{\sin^2\theta_A}\bar{\eta}) \qquad (A4.84)$$

The angular dependence for transfer is similar to that for absorption whereby parallel oscillators have the greatest probability for exchange of the energy and perpendicular oscillators have no probability with the explicit relationship being

$$\text{Probability of energy transfer} \propto \cos^2\beta$$

with β being the angle between the donating emission oscillator and the accepting absorption oscillator. Therefore, for randomly oriented oscillators:

$$\overline{\sin^2\theta_A} = \frac{\int_0^\pi \sin^2\theta \cos^2\theta \sin\theta \, d\theta}{\int_0^\pi \cos^2\theta \sin\theta \, d\theta} \tag{A4.85}$$

(where $\sin\beta \propto$ number of oscillators at angle β to the transferring oscillator and $\cos^2\beta \propto$ probability of exchange). Therefore,

$$\overline{\sin^2\theta_A} = \frac{2/15}{1/3} = 2/5 \tag{A4.86}$$

This remains constant for all solutions of randomly distributed oscillators and Eq. (A4.84) becomes

$$\left(\frac{1}{p} - \frac{1}{3}\right) = \left(\frac{1}{p_0} - \frac{1}{3}\right)\left(1 + \frac{3}{5}\bar{n}\right) \tag{A4.87}$$

Chapter 5

Energy Transfer

5.1 Introduction

An electronically excited molecule may, as we have seen, lose its excitation energy by converting it into light or thermal energy, and it may transfer it to another molecule by collision or complex formation (Chapters 2 and 4). In the latter two processes the transfer of energy is effected by the coupling of electronic orbitals between the two molecules. There exist, however, mechanisms permitting the transfer of excitation energy between molecules separated by distances greater than the molecular diameters. The first recorded observation of transfer of this kind was made by Cario and Franck (1923), who found that a mixture of mercury and thallium vapors when irradiated with light that was adsorbed exclusively by mercury showed the emission spectra of both mercury and thallium. Indirect excitation of thallium by reabsorption of mercury emission was excluded in this experimental setup; the excitation energy must accordingly have been transferred from mercury to thallium through a nonradiative process working at a distance. Förster (1949a, b) established that the fluorescence of trypaflavin in solution was quenched by rhodamine B over a distance of 70 Å, i.e., over a distance several times the molecular dimensions.

A now classic experiment demonstrating efficient transfer of excitation energy between chromophores in a protein (myoglobin) was made by Bücher and Kaspers (1947). Carboxymyoglobin is split by photochemical action releasing CO, when light is absorbed by a heme group. In the presence of oxygen an exchange takes place:

$$MbCO + O_2 \underset{}{\overset{h\nu}{\rightleftarrows}} MbO_2 + CO$$

131

This conversion can be measured spectrophotometrically. The absorption spectra of carboxymyoglobin and its heme and protein parts are plotted in Fig. 5.1, which originates from Bücher and Kaspers (1947). From these spectra the relative absorption by the two main types of chromophores (aromatic amino acids and carboxyheme) can easily be

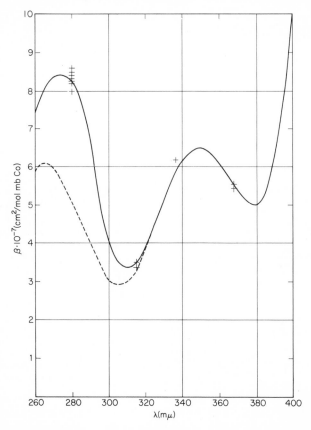

Fig. 5.1. Photochemical action spectrum for carboxymyoglobin (crosses) and absorption spectra of carboxymyoglobin (solid line) and its components, carboxyheme (dashed line) and protein (dotted line) (Bücher and Kaspers, 1947).

calculated. Bücher and Kaspers used monochromatic light of wavelengths from 280 to 546 nm and found a quantum efficiency of one in the whole spectral range; i.e., the quanta absorbed by the aromatic amino acid residues were just as efficient in effecting the splitting off of carbon monoxide as those absorbed directly by heme. This can be explained only if we assume the existence of a very efficient mechanism for transfer of excitation energy over relatively large distances. As we shall see (in Section 5.3), later investigations (Weber and Teale, 1959) have shown that in this particular system the transfer efficiency is actually close to unity. Similarly Stryer (1965) found almost 100% transfer of energy absorbed by tryptophan and tyrosine in apomyoglobin* and apohemoglobin* to ANS† adsorbed (molar ratio to heme sites 1:1) to the protein, the excitation spectrum in each case coinciding with the sum of the absorptions of ANS and the aromatic amino acids.

5.2 Transfer Mechanisms

Being a charged oscillator, an electron excited to a singlet or triplet state in a molecule creates a field extending beyond the limits of the molecule. Onto this field is imposed the frequency characteristic of the excited electron. A mechanical oscillator (e.g., a tuning fork) may transfer kinetic energy through a "field" to a second oscillator so tuned that it may vibrate in resonance with the first one. Similarly, an excited molecule may transfer its excitation energy to a second molecule capable of an electronic transition corresponding to the frequency imposed on the field by the primarily excited oscillator.

We call the two molecules (or separated chromophores in a large molecule) involved in the process the *sensitizer* or *donor*, and the *acceptor*, respectively. One requirement for a transfer of this kind is obviously that the frequency corresponding to the energy gap of a transition from the excited state to the ground state in the donor is the same as the frequency corresponding to the reverse process in the acceptor. The phenomenon is usually referred to as *resonance transfer*. When the acceptor is a fluorescent compound it may also be called *sensitized fluorescence*.

In molecular oscillators (in contrast to free atoms) one electronic transition does not correspond to one fixed frequency, but, due to the various vibrational transitions superimposed, to a whole set of frequencies (Chapter 1). In the condensed phase, which is of primary interest to us here, the spectrum is "smeared out" into a Gaussian-type curve, and a continuous interval of frequencies is imposed on the field by one *type* of electronic transition, and consequently frequencies within a certain distribution may bring about excitation of the acceptor. (Single molecular events are, however, governed by the laws of

*Myoglobin and hemoglobin from which the heme groups have been removed.
†1-Anilinonaphthalene-8-sulfonate.

quantum mechanics, and each individual transfer is a one-step phenomenon, an either-or process taking place at one distinct frequency.)

The conditions necessary for resonance transfer are: (a) the donor
must be a fluorescent (singlet transfer) or phosphorescent (triplet
transfer) group with a sufficiently long lifetime; (b) the absorption
spectrum of the energy acceptor must overlap the emission spectrum of
the donor (Fig. 5.2); (c) the relative orientations of the oscillators must
permit strong interaction; (d) the donor and the acceptor must be within
a certain distance for a given efficiency of transfer (Styrer, 1960).

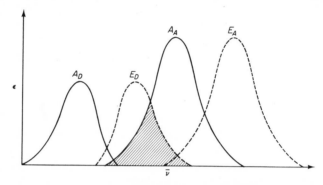

Fig. 5.2. Spectral overlap (hatched area) between donor emission (E_D) and acceptor
absorption (A_A) required for resonance transfer.

Obviously the conditions are similar to those of trivial reabsorption,
and unless special precautions are taken it may in some cases be
difficult to distinguish between the two processes. As we see from
Table 5.1, which is based on a paper by Förster (1959), it is usually
relatively easy to determine the type of mechanism responsible for
transfer or excitation energy in a given system. Further, in most cases
reabsorption is easily avoided by keeping the optical densities low

TABLE 5.1 Characteristic Properties of Transfer Mechanism[a]

	Resonance transfer	Reabsorption	Complexing	Collision
Dependence on sample volume	None	Increase	None	None
Dependence on viscosity	None	None	None	Decreased
Donor lifetime	Decreased	Unchanged	Unchanged	Decreased
Donor fluorescence spectrum	Unchanged	Changed	Unchanged	Unchanged
Absorption spectrum	Unchanged	Unchanged	Changed	Unchanged

[a]From Förster (1959).

enough (which is possible only in studies of *intra*molecular resonance transfer).

Several classes of resonance transfer have been described (cf. Förster, 1960), depending on the strength of the interaction. Very strong interaction (interaction energy ≫ vibrational energies) is usually encountered in crystals and in highly ordered macromolecular systems. Since the transfer process is in those cases much more rapid than the vibrational, rotational, and translational relaxations, we may describe the excitation as being delocalized. In the vast majority of cases where resonance transfer is encountered in fluorescence work, however, the interaction energy is relatively moderate and the transfer process slow in comparison with the collision rates. The interaction is usually between dipoles. In the transfer process these will synchronize, the energy of the combined system thus becoming lower than that of the two components, and there will consequently be an attraction (van der Waal's type) between them. According to experimental observations and the Förster theory (Section 5.3), the upper limit of the distance permitting such interactions is of the order 50–100 Å.

5.3 Quantitative Theory of Energy Transfer

The quantitative aspects of resonance transfer were first outlined by J. and F. Perrin (1924, 1927, 1932), but a complete theory has been developed mainly by Förster (1948). Although it is beyond the scope of this text to elaborate the quantum mechanical background of this theory, we will give an outline of some concepts, and derivations leading to useful mathematical expressions. [For a more detailed discussion see Förster (1948, 1951).]

The field strength at a point A produced by an excited electron oscillating at D is

$$E = \frac{\kappa e a}{n^2 R^3} \qquad (5.1)$$

where e is the charge of the electron, κ is an orientation factor of the magnitude 1, a is the amplitude of the oscillation at D, n is the refractive index, and R is the distance between D (the energy donor) and A (the acceptor). By use of a classic approach it can be demonstrated that, after a time t of exact resonance, oscillator A has acquired the energy

$$W_A = \frac{e^2 E^2}{8m} t^2 \qquad (5.2)$$

where m is the mass of the electron. Consequently, the interaction energy between the oscillators (dipoles) is proportional to the 6th power of the distance [Eq. (5.1)]. The rate of transfer between D and A is

$$-\frac{dW_D}{dt} = \frac{dW_A}{dt} \tag{5.3}$$

Translated into quantum mechanical concepts,

$$\frac{dW_A}{dt} = N_{D \to A} h\nu \tag{5.4}$$

where $N_{D \to A}$ is the number of quanta transferred per second, and $h\nu$ is the energy of each quantum. For practical reasons a unique (so called critical distance R_0 for each donor-acceptor system has been defined:

$$N_{D \to A} = \frac{1}{\tau_0} \left(\frac{R_0}{R} \right)^6 \tag{5.5}$$

τ_0 is here the lifetime of the excited state of D in the absence of transfer, and $1/\tau_0$ the rate at which the excited state is deactivated in the absence of transfer. It is obvious that R_0 is the distance at which there exist equal probabilities for transfer and for intramolecular deactivation of the excited state of D by radiative or nonradiative processes. Considering the fact that in actual systems in the condensed phase (in which the spectra of transitions are broad bands instead of lines), the resonance is not exact, the frequency dependence of an oscillation is described by the oscillator strength, $f = \text{const.} \int_0^\infty \epsilon(\bar\nu) \, d\bar\nu$. Förster has derived the following expression for the critical distance:

$$R_0^6 = \frac{9 \times 10^6 (\ln 10)^2 \kappa^2 c \, \tau_0 J_{\bar\nu}}{16\pi^4 n^2 N^2 \bar\nu_0^2} \tag{5.6}$$

where c is the velocity of light, N is Avagdro's number, $\bar\nu_0$ the arithmetic mean of the wave numbers of maximum absorption and emission of the donor, and $J_{\bar\nu}$ the overlap integral of the emission spectrum of D and the absorption spectrum of A. Since absorption spectra are generally far more accurately recorded than emission spectra, use is preferably made of the observation by Lewschin (1931) that the two spectra for one oscillator are close to mirror images (when plotted on a wave number scale) with a symmetry axis at the wave number corresponding to the energy gap between the thermally relaxed ground and excited states [approximately identical with $\bar{\nu}_0$ in Eq. (5.6)]. Consequently we may use the approximation

$$J_{\bar{\nu}} = \int_0^{\infty} \epsilon_A(\bar{\nu})\epsilon_D(2\bar{\nu}_0 - \bar{\nu})\, d\bar{\nu} \qquad (5.7)$$

for the overlap integral, where $\epsilon_A(\bar{\nu})$ is the molar decadic extinction coefficient of the acceptor at wave number $\bar{\nu}$ and $\epsilon_D(2\bar{\nu}_0 - \bar{\nu})$ that of the donor at wave number $2\bar{\nu}_0 - \bar{\nu}$. This integral is obtained graphically after plotting the two spectra on a wave number scale (see below and Fig. 5.3).

An alternative procedure has been used by Weber (1960), who estimated the overlap integral after assuming the spectral bands to follow approximately Gaussian distributions.

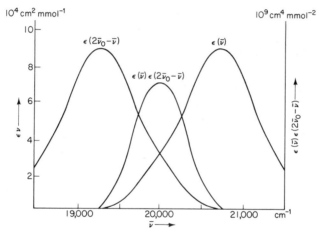

Fig. 5.3. Absorption curve $\epsilon(\bar{\nu})$, its mirror image $\epsilon(2\bar{\nu}_0 - \bar{\nu})$, and the product of the two for fluorescein in water ($10^{-5}\,M$ in $10^{-2}\,M$ NaOH)(Förster, 1948).

κ depends on the angle between the two oscillators, and according to an elementary formula for the interaction energy between two dipoles:

$$\kappa = \cos\phi_{DA} - 3\cos\phi_D \cos\phi_A \qquad (5.8)$$

where ϕ_D and ϕ_A are the angles between the individual oscillators and the line connecting them, and ϕ_{DA} is the angle between the directions of the two oscillations. Because of the (usually) rapid rotation of the oscillators, the statistical average of $\kappa^2 = \frac{2}{3}$ may be used in most calculations.

If in Eq. (5.6) we insert the values $\kappa^2 = \frac{2}{3}$, $c = 3 \times 10^{10}$ cm sec^{-1}, and $N = 6.02 \times 10^{23}$ mol^{-1}, we obtain

$$R_0^6 = \frac{1.69 \times 10^{-33} \tau_0}{n^2 \bar{\nu}_0^2} J_{\bar{\nu}} \tag{5.9}$$

from which R_0 may be calculated.

We shall illustrate this by the example given by Förster (1948) in his earliest complete presentation of the theory, in which he used Eq. (5.9) to calculate R_0 for fluorescein–fluorescein transfer in water. Figure 5.3 depicts the absorption spectrum of fluorescein and its mirror image (i.e., the approximate emission spectrum), and also the product of the two spectra [cf. Eq. (5.7)]. By graphic integration of the latter function one obtains

$$J_{\bar{\nu}} = 1.32 \times 10^{12} \quad \text{cm}^3 \text{ mmol}^{-2}$$

and from the figure we also read $\bar{\nu}_0 = 19{,}900$ cm^{-1}. For this wave number $n = 1.34$ in water, and from earlier measurements it is known $\tau_0 = 5.1 \times 10^{-9}$. Inserting these values in Eq. (5.9) we obtain

$$R_0^6 = \frac{1.69 \times 10^{-33} \cdot 5.1 \times 10^{-9} \cdot 1.32 \times 10^{12}}{1.34^2 \cdot 1.99^2 \times 10^8} = 1.6 \times 10^{-38} \quad \text{cm}^6$$

$$R_0 = 50 \quad \text{Å}$$

Weber and Teale (1959) and Weber (1960) have calculated R_0 values for some transfer pairs and compared these with experimental values. (A method for experimental determination of R_0 by measurements of depolarization of fluorescence has been described by Weber (1954, 1960). (See Chapter 4.) As can be seen from Table 5.2, the agreement is, in general, satisfactory.

TABLE 5.2 R_0 VALUES

	R_0(Å)	
Donor–acceptor pair	Theoretical from (5.9)	Experimental
DNS^a–heme (ferrous)	63	43–67
DNS–heme (ferric)	58	37–69
Phenol–phenol	11	17
Indole–indole	23	16
Phenol–indole	15	20
Tryptophan–DNS	26	24

^{a} 1-Dimethylaminonaphthalene-5-sulfonate.

There is an abundance of experimental data giving support to the validity of the Förster theory. For example, Weber and Teale (1959) conjugated myoglobin with DNS* and studied the transfer efficiencies between this group and heme in the reduced and oxidized states. The valency change of the iron produces large changes in the absorption spectrum of the heme group (Fig. 5.4) and consequently a change in the overlap integral (other variables remaining virtually constant). The overlap integrals calculated were:

ferromyoglobin: 2.8×10^{10} cm³ mmol⁻²

ferrimyoglobin: 1.7×10^{10} cm³ mmol⁻²

the ratio between the two being 1.65. The ratio between the transfer efficiencies in the reduced oxidized states was experimentally found to be 1.74 ± 0.24, thus giving a good demonstration of the proportionality

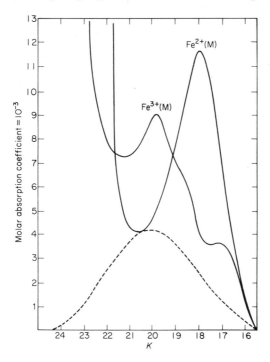

Fig. 5.4. Absorption spectra of ferromyoglobin (Fe²⁺) and ferrimyoglobin (Fe³⁺) and fluorescence spectrum of the DNS conjugate (dashed line). The latter is normalized to a maximum of 4.3×10^3 (molar absorption of corresponding absorption band) (Weber and Teale, 1959).

*1-Dimethylaminonaphthalene-5-sulfonate.

between spectral overlap and transfer efficiency as predicted by the theory.

Let us consider a molecule capable of transferring its excitation energy to a second molecule. By the quantum yield of fluorescence of this molecule we mean the ratio between the number of quanta which are emitted directly as light and the total number of quanta absorbed, i.e., the sum of those emitted as light and those which are dissipated in other ways. In an isolated molecule whose excited state lifetime is τ_0 sec, the total rate at which absorbed quanta are dissipated is $1/\tau_0$ sec^{-1}, and the rate at which quanta are emitted as fluorescence $q_0(1/\tau_0)$ sec^{-1}, where q_0 is the quantum yield. In the presence of a second molecule to which the excitations may be transferred, a new mode of deactivation of the excited state is added. The various processes may be regarded as competing first-order reactions (Fig. 5.5),

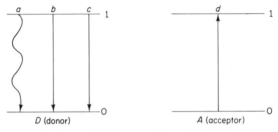

Fig. 5.5. Deactivation and excitation in a pair of oscillators displaying resonance transfer. a = nonradiative intramolecular deactivation $D^1 \rightarrow D^0$, b = fluorescence deactivation $D^1 \rightarrow D^0$, c = nonradiative deactivation $D^1 \rightarrow D^0$ by transfer to A, d = excitation $A^0 \rightarrow A^1$ by transfer from D. Rates (N_x sec^{-1}): $N_a + N_b = 1/\tau_0$ = total rate of intramolecular deactivation $D^1 \rightarrow D^0$; $N_c = N_d = D_{D \rightarrow A}$ = rate of excitation transfer.

and the quantum yield of emission q from the first molecule is in this case clearly proportional to the ratio between the rate of intramolecular inactivation of the excited states:

$$q = q_0 \frac{1/\tau_0}{1/\tau_0 + N_{D \rightarrow A}} \tag{5.10}$$

or, using the expression for $N_{D \rightarrow A}$ from Eq. (5.5),

$$q = q_0 \frac{1}{1 + (R_0/R)^6} \tag{5.11}$$

Correspondingly we obtain for the quantum yield of transfer to the second molecule

$$q_T = \frac{(R_0/R)^6}{1 + (R_0/R)^6} \tag{5.12}$$

In this context we can easily explain the decrease in donor lifetime listed in Table 5.1 as one of the characteristics of resonance transfer. Since transfer of excitation energy to an acceptor molecule increases the rate at which the excited state of the donor is deactivated, the average lifetime of the donor has to decrease with increasing transfer efficiency. It should be noted that the quantum yield of transfer is not limited to the maximum quantum yield of fluorescence of the donor, but may exceed this value by orders of magnitude. An example is the 100% transfer efficiency noted between aromatic amino acids and heme in myoglobin (Section 5.1), whereas the quantum yield of fluorescence is only 20%.

5.4 Practical Applications

We will consider some practical applications of the equations given in the preceding section. Weber (1961) labeled bovine serum albumin with an average of 1.3 DNS groups per protein molecule and studied the effects on excitation transfer from aromatic amino acids to DNS caused by 9.5 M urea and 30% methyl-ethyl carbinol. With the help of Eq. (5.11) and the R_0 values (see Table 5.2) the average increase in interchromophoric distance and the volume change (assuming isotropic expansion) could be calculated (Table 5.3).

TABLE 5.3 EXPANSION OF BOVINE SERUM ALBUMIN

Solvent	$R(\text{Å})$	$R-R_{\text{water}}(\text{Å})$	Volume change
Water	19.5		
30% Methyl-ethyl			
carbinol	23.3	3.8	60%
0.5 M urea	25.8	6.3	90%

Models describing the mechanisms of radiation effects on biological material involve the assumption of migration of absorbed energy in macromolecules. Studying the effects of hydration water on excitation transfer in a solid protein, Rosén and Anhström (1965) used thin films of bovine serum albumin labeled with DNS. The protein films (on quartz plates) were equilibrated with air of various relative humidities (the final water contents were determined gravimetrically on larger protein samples subjected to the same treatment), and absorption and fluorescence spectra were recorded. The transfer efficiencies were calculated from those spectra and also theoretically (normalized to the experimental value corresponding to the lowest water content and assuming isotropic expansion) with the help of Eq. (5.12). The close resemblance between the two curves (Fig. 5.6) is a strong indication that the swelling of the protein constitutes the major physical influence of hydration water on the transfer process.

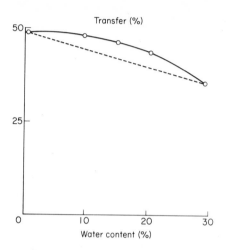

Fig. 5.6. Quantum efficiencies of excitation transfer from aromatic amino acids to DNS conjugated to bovine serum albumin (molar ratio 2.5) at various water contents. Circles: experimental values. Dashed line: calculated from expansion (Rosén and Ahnström, 1965).

We have already mentioned the work of Weber and Teale (1959), in which the investigators studied the quenching effects of heme groups in a variety of heme proteins on the fluorescence of tryptophan (in the native proteins) and of DNS (in conjugates). Computations based on Eq. (5.11) yielded information about the positions of the heme groups not accessible with any other technique before the detailed structures of the proteins had been elucidated by X-ray crystallography. It was also possible to confirm in these experiments the conclusions of Bücher and Kaspers (Section 5.1) that transfer from aromatic amino acids to heme in myoglobin may occur with approximately 100% efficiency. The absolute quantum yields of fluorescence of the heme-free globins were found to be about 20%, which means that radiationless transitions from the singlet excited state are 4 or 5 times more probable than fluorescence:

$$q_0 = \frac{N_b}{N_a + N_b} \tag{5.13}$$

(using the symbols introduced in Fig. 5.5). On the other hand, from fluorescence measurements before and after removal of the heme, it was found that transfer of the excited state to the heme has at least 100 times greater probability than fluorescence, and therefore at least 20 times greater than the other competing radiationless processes.

By use of a geometrically well-defined system (in contrast to earlier work in which various distributions of the chromophores had to be assumed) Latt *et al.* (1965) have demonstrated that intramolecular dimensions may actually be determined within an error of about ±10% using the quantitative theory of resonance transfer outlined. The system used was a (rigid) synthetic, decacyclic steroid, to the opposite ends of which two different fluorescent groups were attached. The distances between the chromophores were measured on molecular models and compared with those calculated from excitation transfer measurements (5.9) and (5.12).

Weber and Daniel (1966) have studied the binding of ANS to bovine serum albumin. Calculations of energy transfer from tryptophan to ligand and from ligand to ligand allowed elucidation of the distribution of the ligands. The potentialities of energy transfer studies are demonstrated in this work, from which Fig. 5.7 has been taken. This figure gives a good illustration of how, due to tryptophan-ligand transfer, an increase of the ratio of ligand to protein results in a decrease of the protein fluorescence and an increase of the ligand fluorescence. For the single point at which all the curves cross, the authors introduced the term *isoemissive point*. The significance of this feature is in every way the same as that of an isosbestic point in absorption spectroscopy.

5.5 Concentration Depolarization

As we have seen above, resonance transfer may be studied qualitatively and quantitatively either by observing the quenching of the donor effected by transfer to the acceptor, or by observing the increase of acceptor fluorescence (sensitized fluorescence) induced in the same way. These methods are applicable in the study of transfer between unlike molecules, for which the spectra can be recorded individually, but in the study of transfer between like molecules this is obviously not possible, and an entirely different technique has been developed to follow such transfer, utilizing fluorescence polarization measurements. Energy as a depolarizing factor has been given some consideration in Chapter 4. We will further elaborate the quantitative relationships pertaining to this effect and discuss some experimental consequences.

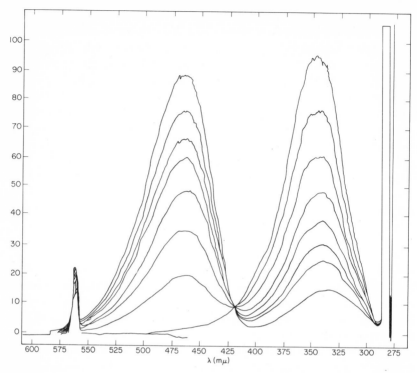

Fig. 5.7. Spectra of mixtures of ANS and bovine serum albumin at ratios of ligand/ protein of 0 to 2. The added ANS is stoichiometrically bound to the protein, since the concentrations ·of both are much higher than the dissociation constant (Daniel and Weber, 1966).

When the concentration of a fluorescent compound in solution is increased, eventually a point is reached where a decrease in the quantum yield becomes noticeable (see Chapter 2). At a lower concentration (sometimes much lower), an effect of the interaction between solute molecules may be observed, which is a drop in the polarization of the fluorescence. This phenomenon is referred to as *concentration depolarization*. A graphic illustration is given in Fig. 5.8. As discussed in Chapter 4, the reason for this depolarization, which takes place even in a solvent of infinite viscosity, is that the absorbed quantum of light is transferred from one oscillator to another, with a certain angle between them, before it is reemitted as radiation. One or more steps may be involved. The transfer may occur in two different ways: radiative or nonradiative (resonance transfer). The first type, which we can eliminate in experimental measurements (Table 5.1), is of no interest, and we will not discuss it further.

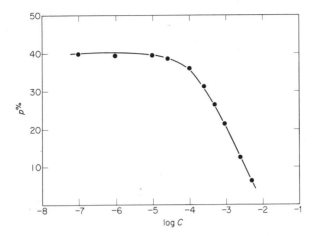

Fig. 5.8. Concentration depolarization of fluorescein in glycerine (Pheofilov and Sveshnikov, 1940).

Concentration depolarization has received much attention for a long time, and Pheofilov and Sveshnikov (1940) have shown that a simple empirical relation exists between the polarization and the concentration:

$$\frac{1}{p} = \frac{1}{p_0} + Ac\tau_0 \qquad (5.14)$$

where p is the polarization at concentration c of the fluorescent species, p_0 is the polarization at infinite dilution, τ_0 is the lifetime of the excited state (which does not change until a concentration is reached where concentration quenching takes place), and A is a constant. A full derivation of this relation for a solvent of infinite viscosity, which applies to the whole concentration range, and which is expressed in terms of the average number of transfers between like oscillators, has been presented by Weber (1954). This derivation, which need not be reproduced here, yields

$$\frac{1}{p} \pm \frac{1}{3} = \left(\frac{1}{p_0} \pm \frac{1}{3}\right)\left(1 + \frac{4N\pi cR_0^6}{15(2a)^3} \times 10^{-3}\right) \qquad (5.15)$$

where the plus signs apply to excitation with natural (unpolarized) light and the minus signs to excitation with completely polarized light (excitation and emission at right angles). $c =$ concentration in mole/liter, $R_0 =$ critical distance, $a =$ effective molecular radius, and p and p_0 are defined as in Eq. (5.14). In accordance with (5.5), R_0^6 may be replaced by $N_{D\rightarrow A}R_{\tau_0}^6$, and the identity between (5.14) and (5.15) becomes obvious.

This linear function may be utilized to compute the value of R_0, provided a proper value of a is assumed ($\sim 10^{-7}$ cm). From the slope s of the straight line obtained when $1/p \pm \frac{1}{3}$ is plotted aganist c, we get

$$R_0 = (2a)^{1/2}\left[\frac{15s \times 10^3}{4N\pi\left(\frac{1}{p_0}\pm\frac{1}{3}\right)}\right]^{1/6}.$$ (5.16)

This equation has been successfully applied in the determination of R_0 values, e.g., by Weber for protein chromophores (see Section 5.3).

The linear dependence upon concentration, (5.14) and (5.15), may be explained/understood intuitively in the following way. Picture an excited molecule D, which is capable of transferring excitation energy to another molecule A, which may or may not be similar to D. A is located within a spherical shell, the inner radius of which equals the sum of the radii of D and A (in the case where D and A are identical molecules, the radius equals $2a$ as in (5.15) and (5.16), and the outer radius is dependent upon R_0, the critical distance). The probability of a molecule A being within this spherical shell surrounding D is then directly proportional to the concentration of A. Clearly it is possible to define a critical concentration in analogy with the critical distance R_0 such that

$$c_0 \propto \frac{1}{(R_0)^3}$$ (5.17)

The exact relation between these entities may be formulated in a number of ways; we will here present the one given by Förster (1951).

According to this definition the critical concentration c_0 is defined in a static system with randomly distributed and oriented molecules as the concentration at which the probability of the presence of a molecule A within a sphere with radius R_0 equals $1 - 1/e = 0.63$. Simple geometrical considerations then yield

$$c_0 = \frac{3}{4N\pi\bar{R}_0^3} = \left[\frac{7.35 \times 10^{-8}}{\bar{R}_0}\right]^3$$ (5.18)

For fluorescent dyes having relatively high quantum yields of fluorescence and relatively extensive overlap between their absorption and emission spectra, c_0 is of the order of 10^{-3}, but it may be several orders of magnitude higher for chromophores not meeting these criteria.

Although concentration depolarization has until now been mainly a technique for study of simple systems such as independent, fluorescent molecules in frozen solutions, e.g., in the determination of R_0 values, it has a potential use in more direct biological investigations. The term concentration depolarization suggests a phenomenon occurring between solute molecules in a solution of sufficient concentration, but basically the situation will be very similar for any system in which two or more molecules of a fluorescent compound are brought close enough together for transfer to take place. A more general term would be fluorescence depolarization through energy transfer. The type of system which will be of primary interest is one in which two or more fluorescent groups (natural, such as coenzyme, or various synthetic groups which have been employed in such studies) are bound to one and the same macromolecule or aggregate of macromolecules. Provided the interchromophoric distances are of the order of, or smaller than, the critical distance, and the oscillators are not at right angles to each other, transfer will take place between the groups, which will be observed as decreased polarization. Depolarization studies in such systems may be a more powerful tool than energy transfer calculations based on observed variations in quantum yields (Section 5.4) for two reasons: (1) applicability to systems in which transfer takes place between identical molecules, and (2) higher sensitivity to variations in angular distributions of the chromophores. Depolarization studies may thus yield information regarding both angular and spatial distributions of chromophores.

Semiquantitative studies of depolarization through energy transfer include simple techniques, e.g., in studies of protein-ligand interactions (Weber and Daniel, 1966; Weber and Young, 1964), but a rigorous mathematical treatment of a system which consists of a macromolecule and more than two chromophores which take part in the transfer tends to become very complex, and a number of limiting assumptions have to be introduced. (In a system in which we want to study binding to a macromolecule or conformation changes of the latter we may, for instance, have to assume random angular and spatial distributions of the chromophores, which may or may not be a good approximation.) Attempts made in this direction (Weber and Daniel, 1966), however, indicate that a combination of depolarization measurements and more direct energy transfer studies, such as those described in the preceding section, may open new angles of attack on many problems related to biological functions of protein molecules.

148   5   Energy Transfer

Bücher, T., and Kaspers, J. (1947). *Biochim. Biophys. Acta*, **1**, 21.
Cario, G., and Franck, J. (1923). *Z. Physik*, **17**, 202.
Daniel, E., and Weber, G. (1966). *Biochemistry*, **5**, 1893.
Förster, T. (1948). *Ann. Physik. Lpz.*, **2**, 55.
Förster, T. (1949a). *Z. Elektrochemie*, **53**, 93.
Förster, T. (1949b). *X. Naturforschung*, **4a**, 321.
Förster, T. (1951). *Fluoreszenz Organischer Verbindungen*, Vandenhoek und Rup-
recht, Gottingen, Chapter 4.
Förster, T. (1959). *Disc. Faraday Soc.*, **27**, 7.
Förster, T. (1960). *Radiation Res. Suppl.*, **2**, 326.
Latt, S. A., Cheung, H. T., and Blout, E. R. (1965). *J. Amer. Chem. Soc.*, **87**, 995.
Lewschin, W. L., *Z. Physik* (1931). **72**, 368.
Perrin, F. (1932). *Ann. Chim. Physique*, **17**, 283.
Perrin, J. (1925). *2me counseil de Chemie Solvay, Bruxelles*, Gauther–Villars, Paris,
p. 322.
Perrin, J. (1927). *Compt. Rend.*, **184**, 1097.
Pheofilov, P. P., and Sveshnikov, B., (1940). *J. Physics USSR*, **3**, 493.
Rosen, C.-G., and Ahnström, G. (1965). *Intern J. Radiation Biol.*, **9**, 435.
Stryer, L. (1960). *Radiation Res. Suppl.*, **2**, 432.
Stryer, L. (1965). *J. Mol. Biol.*, **13**, 482.
Weber, G. (1954). *Trans. Faraday Soc.*, **50**, 552.
Weber, G. (1960). *Biochem. J.*, **75**, 335.
Weber, G. (1961). In *Symposium on Light and Life*, The Johns Hopkins Press, Balti-
more, p. 82.
Weber, G., and Daniel, E. (1966). *Biochemistry*, **5**, 1900.
Weber, G., and Teale, F. W. J. (1959). *Disc. Faraday Soc.*, **27**, 134.
Weber, G., and Young, L. B. (1964). *J. Biol. Chem.*, **239**, 1415.

Chapter 6

Quantitative Aspects of Fluorescence Measurements

6.1 Introduction

Any discussion concerning the quantitative measurement of the fluorescence parameters of a solution must concern itself with the methods and instrumentation used. To completely describe the fluorescence of a compound in solution it is essential to measure the amount of light energy absorbed, the excitation spectrum, the emission spectrum, the quantum yield, the number of components in the solution, the lifetime of the excited state, and the polarization of the emitted fluorescence. This chapter will deal with the problem of quantitation from both the theoretical and instrumental points of view. One must understand the limitations on what can be measured in order to understand why certain equations are used in analysis. It may not appear obvious why such complicated equations are derived, but it becomes very clear when the limitations of instrumentation are seen. There are many difficulties in measuring fluorescence quantities. Corrections must be introduced as a result of the specific physical instrumentation employed as well as the particular fluor to be examined. As stated in Section 2.5, many of these fluors will dimerize or undergo spectral changes with concentration changes; similarly, photodecomposition will occur with some fluors during the time taken to make the measurements if the exciting light is too intense.

In this chapter methods of calibration will be described which afford a reasonable degree of accuracy for quantitative measurement of fluorescence parameters, particularly the quantum yield. It may be emphasized that the values obtained for quantities such as the quantum yield of various compounds have differed widely from laboratory to laboratory for a number of years. It is thus not as simple as it appears to get accurate values of fluorescence parameters. Currently, the most difficult value to obtain is that of the lifetime of the excited state. Recent advances in instrumentation will hopefully resolve some of the basic disagreements between laboratories. The reader is assumed to be familiar with the basic instrumental components used in spectroscopy; if not, the following books are recommended: *Chemical Instrumentation* by H. A. Strobel, Addison-Wesley, Reading, Mass., 1960 and *Undergraduate Instrumental Analysis* by J. W. Robinson, Marcel Dekker, New York, 1970.

6.2 Relation between Absorption and Fluorescence

A. ABSORPTION

The electronic events connected with absorption of electromagnetic radiation have been discussed in Chapter 1. What is of significance in quantitative spectroscopy, however, is the relationship between the intensity of light incident upon an absorbing object and the emergent intensity, for the ratio of the two intensities is a measure of the amount of light absorbed (see Fig. 6.1.)

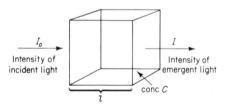

I_0
Intensity of
incident light

I
Intensity of
emergent light

conc C

l

Fig. 6.1. Schematic representation of incident and emergent light.

The Lambert-Beer law described in Sec. 1.4 states the relationship of these intensities

$$I = I_0 \exp\left(-\epsilon c l\right) \qquad (6.1)$$

One must remember that the extinction coefficient and the measurements of intensity are at specific wavelengths. The equation might be more exactly written:

$$A = \log_e \frac{I_0(\lambda)}{I(\lambda)} = c\, l\epsilon(\lambda) \qquad (6.2)$$

Spectrophotometers which measure absorbancy are generally constructed according to the schematic diagram shown in Fig. 6.2. Notice that arrangements are made to select the wavelength (λ) of light incident upon the sample but no provision to insure that only the same

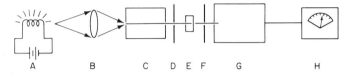

A   B   C   D E F   G   H

Fig. 6.2. Schematic of single beam spectrophotometers. A = light source: (1) tungsten filament for near ultraviolet and visible and (2) hydrogen discharge for ultraviolet; B = collimating and focusing lens; C = wavelength selector: (1) filters, and (2) grating or prism monochromator; D = adjustable slit to allow variable bandwidth and variable intensity; E = sample cuvette; F = collimating slit (usually not adjustable); G = detector; barrier cell, thermopile, or photomultiplier tube; H = galvanometer to measure signal from G.

wavelength of light is picked up by the detector. This normally has no significance, except when the sample is fluorescent. In this case, there is a contribution to the $I(\lambda)$ (in 6.3) by the fluorescent light, since

$$I(\lambda) \text{ (at the detector)} = I(\lambda) + \alpha I(\lambda_f) \qquad (6.3)$$

where $I(\lambda)$ is the same quantity as described in Eq. (6.2). $I(\lambda_f)$ is the intensity of fluorescent light incident on the detector, α is a proportionality factor to account for the different sensitivity of the detector for λ_f as opposed to λ. This results in a net increase in the apparent $I(\lambda)$ and therefore gives an erroneously low absorbancy reading. The actual effect of this phenomenon is usually quite small since fluorescence is a radially dispersive phenomenon and therefore the amount of light reaching the detector is inversely proportional to the square of the distance between the detector and the sample cuvette; whereas, the transversing beam is directed toward the detector.

A sample test for the above simply involves manually placing the suspected fluorescent sample at different distances and noting if any change in the absorbancy readings occurs. If the absorbancy decreases as the sample nears the detector, then fluorescence of the sample is registering on the detector. This fluorescent component is being recorded and may be corrected by placing a filter between the sample and detector which has high transmission for the exciting light but very low transmission for fluorescent light.

B. FLUORESCENCE

Fluorescence is intimately related to the process of absorption, as discussed in Chapters 1–3. There are two characteristics of fluorescence that make the quantitation quite different in terms of instrumentation and expression from that of absorption spectroscopy; the radially dispersive nature of the fluorescence emission and the "Stokes shift" of the emission spectrum.

Experimentally, the first characteristic means that unless some type of integrating sphere is used (see Sec. 6.5B) only a small fraction of the total fluorescent light will be detected (see Fig. 6.3 for schematic of

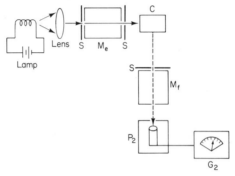

Fig. 6.3. A right angle spectrofluorometer. S = variable slit; M_e = excitation monochromator; C = sample cell; M_f = fluorescent monochromator for scanning the fluorescence spectra; if one wishes only to measure intensities, a filter excluding exciting light but passing fluorescent light may be used; P_2 = photocell; G_2 = galvanometer to show output from P_2.

a spectrofluorometer). The second characteristic means that fluorescence occurs at longer wavelengths and lower energies than the absorbed light. Thus, the energy intensity of absorption will always be greater than the fluorescent energy intensity even when there is one photon of light emitted for every photon absorbed. This necessitates that all intensity measurements in fluorescence be in units of quanta rather than energy. (Either units may be used for absorption.)

With the above points in mind, the following equation is introduced:

$$F = I_0(\lambda)[1 - 10^{-\epsilon_\lambda cl}][q] \qquad (6.4)$$

where

F = total fluorescence intensity in quanta/second

$I_0(\lambda)$ = intensity of exciting light in quanta/second

$\epsilon_\lambda cl$ = as for Eq. (6.1)

q = quantum efficiency of fluorescence

= quanta emitted/quanta absorbed.

When a very small fraction of light is absorbed ($\epsilon_\lambda cl \lesssim 0.05$) it reduces to

$$F = I_0(\lambda)[2.3\epsilon_\lambda cl][q] \qquad (6.5)$$

As has been mentioned, in practice only a small portion of the emitted light, often at a right angle to the path of the exciting light (as in Fig. 6.3), is measured. This is, however, an acceptable procedure since the total intensity of light measured over a specific solid angle is a proportional representation of the total intensity of emitted light.

6.3 Instrument Calibration

A. Light Source and Excitation Monochromator Efficiency: Intensity vs. Frequency of Light

As will be discussed in detail, in order to obtain a true fluorescence excitation spectrum, it is necessary to know the energy or quantum distribution of a light source and monochromator as a function of frequency (or wavelength) of the exciting light. There are two common methods of obtaining this information, viz calibration with a thermopile or a photon counter.

In the first instance, a thermopile is placed in line with the beam from the excitation monochromator [Fig. 6.4(a)] and the thermopile output is recorded as a function of wave number. Since the thermopile records relative intensities as energy, these readings must be divided by the frequency (or multiplied by the wavelength) to obtain relative intensities in units of quanta, remembering that $E = h\nu$, see Chapter 1.

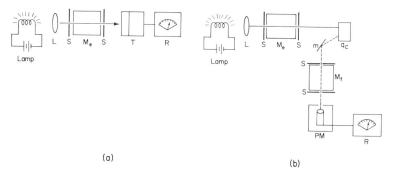

(a)   (b)

Fig. 6.4. (a) L = lens; S = slit (variable); M_e = excitation monochromator; T = sensitive thermopile; R = output recorder. (b) L, S, M_e, and R as in (a); q_c = quantum counter; m = polished mirror; M_f = emission monochromator; PM = photomultiplier.

It is the choice of the investigator whether to plot relative intensities against wavelength or frequency, the choice is dependent upon which of the two units the investigator uses when recording absorption and fluorescence spectra. Some have preferred plotting against frequency (or wave number) since absorption and fluorescence bands are symmetrical with frequency and not with wavelength, thus allowing a more accurate representation of the mirror image symmetry between the fluorescence spectrum and the last absorption band (see Chapters 2 and 3). Other investigators prefer plotting against wavelength since many spectrophotometers are calibrated in wavelength in the visible and ultraviolet regions.

Figure 6.5(a) shows a light intensity curve for a xenon lamp obtained by using a sensitive thermopile. The units of light source output are

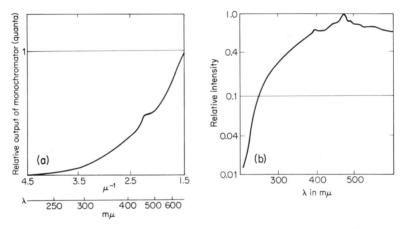

Fig. 6.5. (a) Relative light intensity delivered by a quartz monochromator using a xenon lamp as a source and a sensitive thermopile to measure output (Parker and Rees, 1960). (b) Relative light intensity delivered by a grating monochromator using a xenon lamp as a source and a quantum counter to measure output (Melhuish, 1962).

relative and proportional to number of photons per unit time, not energy (Parker and Rees, 1960).

Figure 6.4(b) shows one of the several possible physical arrangements for measuring lamp intensity by a quantum counter. In this case, the monochromator–photomultiplier detection system records output in quanta, so no recalculation need be made and the output may be plotted directly against frequency or wavelength of the exciting light. Figure 6.5(b) shows a relative light intensity curve for a xenon lamp using a quantum counter (Melhuish, 1962).

The quantum counter indicated in Fig. 6.4(b) is simply a fluorescent solution of sufficient concentration to have very nearly 100% absorption (infinite absorbancy) in the range of exciting frequencies to be studied. Furthermore, the fluorescent substance must have a quantum yield of fluorescence which is independent of the exciting wavelength, a characteristic shown by the majority of low molecular weight fluorescent materials (see Sec. 2.5). Three compounds that are readily available, easy to purify, and useful as quantum counters are rhodamine B (in ethylene glycol, 3 mg/ml $\sim 6 \times 10^{-3} M$), fluorescein (in $N/10$ NaOH, 2 mg/ml $\sim 10^{-3} M$), and quinine bisulfate (in $1 N$ H_2SO_4, 4 mg/ml $\sim 5 \times 10^{-3} M$).

Only the relative intensities are measured. The monochromator–photomultiplier is set at a prescribed frequency (or wavelength), usually the maximum frequency of fluorescence for the particular quantum counter used, and the relative intensity recorded. From the discussion of Sec. 2.5 it is clear that the fluorescence spectrum of a compound is independent of exciting light; the peak fluorescence is proportional to the total integrated fluorescence, therefore obviating the necessity for obtaining the entire emission spectrum and integrating the area to determine total number of quanta reflected into the detector system.

Two further methods of calibration will be mentioned but not discussed: (1) a calibrated phototube whose spectral sensitivity has been supplied by the manufacturer, and (2) a chemical actinometer (Parker, 1953; Hatchard and Parker, 1956).

B. MONOCHROMATOR AND PHOTOMULTIPLIER EFFICIENCY: RESPONSE VS. FREQUENCY OF THE LIGHT

A fluorescence emission spectrum is obtained by exciting a solution with light of a constant frequency and intensity, while scanning the quantum output of fluorescence with emission monochromator and photomultiplier tube. Thus, if the efficiency of the photomultiplier and monochromator varies with the frequency of emitted light selected by the monochromator, the recorded emission spectrum will be in error according to the change in efficiency (or sensitivity) of the fluorescence detection system. It is necessary to determine the sensitivity curve of this system for a particular instrument to correct the apparent spectra to the true emission spectra.

In general, the sensitivity of a monochromator–photomultiplier can be expressed as follows:

$$S_{\bar{\nu}} = P_{\bar{\nu}} f_{\bar{\nu}} W_{\bar{\nu}} \qquad (6.6)$$

where

> $S_{\bar{\nu}}$ = sensitivity at wave number $\bar{\nu}$
> $P_{\bar{\nu}}$ = (output/quanta) of photomultiplier at wave number $\bar{\nu}$
> $f_{\bar{\nu}}$ = fraction of light transmitted by monochromator at $\bar{\nu}$
> $W_{\bar{\nu}}$ = bandwidth in wave numbers units at $\bar{\nu}$

Equation (6.6) is usually used to calculate the sensitivity of the monochromator–photomultiplier system (henceforth referred to as spectrometer) when the range of frequencies is in the ultraviolet region. The approximate sensitivity curve for a photomultiplier is usually supplied by the manufacturer and can be checked against a thermopile of chemical actinometer. This curve is almost always in terms of energy rather than quanta and must be converted to quanta, either by multiplying by λ or dividing by $\bar{\nu}$.

The bandwidth versus slit width is supplied for a particular monochromator by its manufacturer, the bandwidth per millimeter slit width is reasonably constant for grating monochromators and variable for prism monochromators. However, since the bandwidth is usually given as a function of wavelength, it must be converted to bandwidth in wave numbers. The efficiency of transmission of the monochromator is also supplied by the manufacturer. For reasonably small regions of the spectrum the transmission characteristics are relatively constant, probably varying by less than a factor of 2 for most good prism and grating devices over a 100-nm range. Therefore with little error this transmission factor may be included in the $P_{\bar{\nu}}$ factor.

With the above consideration in mind, we have

$$S = Pw_{\lambda}/\lambda^3 \qquad (6.7)$$

where

> P = energy sensitivity of photomultiplier
> w_{λ} = bandwidth of monochromator in wavelength units at λ
> λ^3 = factor to convert P from energy to quanta and to convert w_{λ} from wavelength units to frequency units.

Another method of deriving the sensitivity of the spectrometer is to use a lamp of known spectral distribution. The one usually used has a tungsten filament of known color temperature [determined by the use of an optical pyrometer (Kostkowski and Lee, 1962)]. The spectrum given by the spectrometer may be compared with the calculated spectral distribution of the tungsten lamp as given by Wien's law, which is a good approximation of Planck's expression for blackbody radiation in the visible and ultraviolet regions and gives easier calculations. This method cannot be used effectively at wavelengths shorter than the near-ultraviolet region because the intensity of the tungsten light falls off very rapidly in this part of the spectrum, causing losses in output and resultant errors in calculation of the spectrometer sensitivity. Stair et al. (1963) have described a quartz–iodine lamp with a coiled-coil filament, which has a calibrated output down to 250 nm, but for the shortest wavelength region (far-ultraviolet) it is necessary to use another method [see Eq. (6.11)].

To obtain the spectrometer sensitivity using a lamp of known spectral distribution the following steps are to be followed. The lamp is placed at the entrance slit of the spectrometer and the response of the photomultiplier is recorded as a function of wave number, such that

$$S_{\bar{\nu}} = \frac{R_L}{(dq/d\nu)_L} \tag{6.8}$$

where

$S_{\bar{\nu}}$ = sensitivity of spectrometer at wave number $\bar{\nu} = 1/\lambda$

R_L = response of photomultiplier to calibrated lamp at wave number $\bar{\nu}$

$(dq/d\bar{\nu})_L$ = spectral distribution in quanta of the calibrated lamp at wave number $\bar{\nu}$

R_L is the directly recorded photocurrent of the photomultiplier. It is possible to obtain lamps that have been calibrated to run a known color temperature [National Bureau of Standards; see Kostkowski and Lee (1962)] and the spectral distribution curves are provided in energy units per wavelength interval, i.e., $(dE/d\lambda)_L$; this must be converted to units of quanta per frequency interval for use in Eq. (6.8):

$$\frac{dq}{d\bar{\nu}} = \lambda^3 \frac{dE}{d\lambda_L} \tag{6.9}$$

The same method may be applied even when a lamp is calibrated by the use of an optical pyrometer, since by this method the spectral distribution is in terms of energy per wavelength interval.

However, if an investigator wishes to calibrate his instrument for wavelength rather than wave number, the following holds:

$$S = \frac{R_L(\lambda)}{(dq/d\lambda)_L} \tag{6.10}$$

and

$$S = \lambda \frac{R_L(\lambda)}{(dE/d\lambda)_L} \qquad (6.11)$$

where

$R_L(\lambda)$ = response of photomultiplier at wavelength λ

$(dE/d\lambda)_L$ = spectral distribution of lamp in energy per unit wavelength interval applicable to the wavelength in question

$(dq/d\lambda)_L = \lambda(dE/d\lambda)_L$ = actual relative number of quanta at wavelength λ

Figure 6.6 shows the curves obtained by Parker and Rees (1960) for the calibration of their spectrometer using lamps of known color temperature.

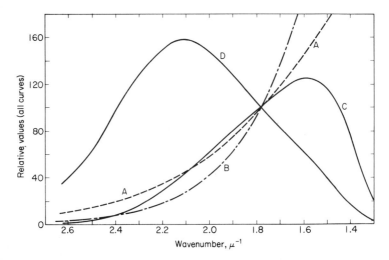

Fig. 6.6. Calculation of sensitivity curve for the visible region (combination of glass-prism spectrometer and E.M.I. 9558 red-sensitive photomultiplier): curve A, relative spectral distribution of light from standard lamp run at a color temperature of 2856°K (energy per unit wavelength interval—$dE/d\lambda$); curve B, relative spectral distribution of light from standard lamp run at a color temperature of 2856°K (quanta per unit frequency interval—$dQ/d\nu$); curve C, photomultiplier output at constant slit width; curve D, sensitivity curve, $(S_{\bar{\nu}})$ (ordinates of curve C divided by ordinates of curve B). All curves normalized to 0.50 μ (Parker and Rees, 1960).

The following references are excellent primary sources for calibration of instruments: Gilmore et al. (1952), Weber and Teale (1957), Parker and Rees (1960), White et al. (1960), and Melhuish (1962). Parker and Rees (1960) have been most often used for representation of results simply because they give their data in graphical form. More involved correction factors may be found in some of the above references for special systems used by the authors: i.e., Gilmore et al. (1952 and 1955) and Shepp (1956) discuss the correction due to a change in refractive index that is necessary when viewing the fluorescence of a solution at an angle other than the normal to the cuvette face.

6.4 Fluorescence Emission Spectra

A. Observed and Corrected Spectra

The observed fluorescence emission spectrum is a function not only of emitted light but also of the transmittance of the fluorescence monochromator as a function of the wavelength of emitted light; in addition, the sensitivity of the photomultiplier is not constant with wavelength. Because of the latter two considerations the overall sensitivity of the spectrometer must be determined as previously described in Sec. 6.3B, so that the observed spectrum may be corrected to give a true emission spectrum.

A source of instability in recording the observed emission spectrum is the fluctuation of intensity of the exciting light. This fluctuation appears in the recordings because the intensity of fluorescence is directly proportional to the exciting light intensity. Parker (1958) has described an instrument in which a change in exciting light intensity is corrected internally by continuously monitoring the exciting light beam with a beam splitter, fluorescent screen, and photomultiplier fed into one arm of a ratio recorder. A similar system has also been described by Weber and Young (1964).

Table 6.1 is a correction curve obtained by Knopp (1967) for the spectrometer portion of the instrument described by Weber and Young (1964). Figure 6.7 shows an observed spectrum for quinine bisulfate in 0.1 N H_2SO_4 obtained from the instrument and the same spectrum corrected according to Table 6.1. Figure 6.7 is the same corrected spectrum for quinine bisulfate only plotted versus wave number of fluorescence rather than wavelength.

B. Noninstrumental Errors

The following is a list of some of the errors that may be encountered during the course of obtaining a fluorescence emission spectrum. Some

TABLE 6.1 RELATIVE EFFICIENCIES OF 6255 PM AND MONOCHROMATOR VIA TUNGSTEN BULB

Wavelength (nm)	Efficiency	Wavelength (nm)	Efficiency	Wavelength (nm)	Efficiency	Wavelength (nm)	Efficiency
300	46.8	400	96.2	500	49.2	600	5.44
305	53.4	405	93.2	505	46.6	605	3.30
310	59.4	410	89.5	510	44.2	610	
315	67.3	415	87.7	515	42.2	615	
320	73.2	420	87.1	520	40.3	620	2.05
325	79.5	425	85.3	525	37.5	625	1.66
330	86.8	430	83.4	530	34.6	630	1.39
335	90.5	435	81.3	535	31.6	635	1.25
340	92.7	440	78.7	540	28.7	640	1.10
345	96.6	445	76.2	545	26.0	645	0.86
350	99.6	450	73.9	550	23.4	650	0.64
355	103.4	455	71.0	555	21.4	655	0.52
360	104.2	460	68.5	560	19.2	660	0.42
365	103.0	465	66.4	565	17.2	665	0.34
370	100.5	470	64.3	570	15.2	670	0.29
375	98.7	475	62.0	575	13.4	675	0.26
380	99.0	480	59.4	580	11.7	680	0.22
385	99.2	485	56.9	585	10.0	685	0.20
390	98.9	490	54.3	590	8.42	690	0.18
395	98.3	495	51.8	595	6.88	695	0.18
						700	0.16

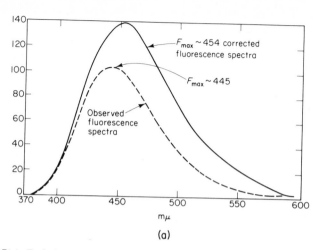

Fig. 6.7(a). Emission spectra of quinine bisulfate plotted against wavelength (mμ).

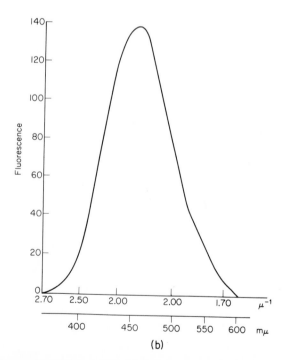

Fig. 6.7(b). Corrected emission spectra of quinine bisulfate plotted against wave numbers (μ^{-1}).

of these artifacts result in a distortion of the spectrum and some may affect only the intensity of the fluorescence. Each source of error will be discussed briefly.

1. Extraneous Fluorescence

This is fluorescence which originates from the quartz or glass cuvettes used to hold the sample being investigated. It is generally advisable to use synthetic quartz cuvettes rather than fused quartz cuvettes, because the former have much less contaminating fluorescence. In addition, when low fluorescence intensities are being measured at high instrument sensitivity, a blank spectrum must be obtained in which the cuvette contains the solvent being used without the fluorescent dye.

Furthermore, in filter fluorometers when working with ultraviolet exciting light, it must be remembered that the filters themselves may be fluorescent. This may be checked again by the use of a "blank" spectrum and can easily be eliminated by placing a sodium nitrite (2.0 M, about 3-mm path length) filter between the cuvette and filter, so as to absorb the scattered exciting light.

2. Reabsorption of Emitted Light

This phenomenon distorts the emission spectrum in the region of wavelengths nearest the absorption band and can occur in those compounds in which the fluorescence and the absorption spectra overlap. This complication may be more easily described by taking the specific example of the dye rhodamine B. Figure 6.8 shows the

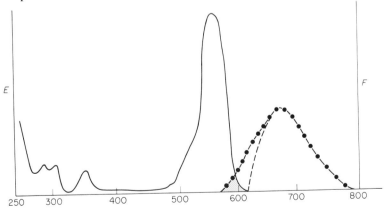

Fig. 6.8. ———, absorption spectrum of Rhodamine B in 1, 2 propanediol; —·—·—, fluorescence spectrum of Rhodamine B in 1, 2 propanediol; ————, theoretically shifted emission spectrum of rhodamine B due to reabsorption of emitted light.

absorption and emission spectra of rhodamine B. The overap interval between the two is shaded. The short wavelength emitted light can be reabsorbed by the dye molecules. At high dye concentrations this reabsorption can become quite significant and results in both a decrease in the measured quantum efficiency* and a shift in the emission spectrum of the dye solution. The emission spectrum will rise more sharply (be less broad) in concentrated dye solutions than in the dilute ones (see Fig. 6.8). These changes are a direct function of dye concentration and will disappear at low dye concentrations.

3. Internal Screening

Often an investigator may excite a dye at a wavelength at which some other nonfluorescent substance in the solution is capable of absorption. The actual light intensity impinging upon the dye is less than the incident light intensity in proportion to the fraction of light absorbed by the nonfluorescent impurity. This will not change the shape of the emission spectrum but will decrease the intensity of the fluorescence, resulting in a need for correction when determining quantum yields of fluorescence (see Sec. 6.8B4).

Equations (6.4) and (6.5) show that the fluorescence intensity is proportional to both exciting light intensity and the absorbance of the substance at this wavelength. Thus, the presence of some absorbing impurity lowers the intensity impinging upon the fluorescent dye and causes the emitted intensity to be lower than expected for the measured exciting light intensity and absorbancy of the solution. This is approximately corrected by multiplying the measured fluorescence intensity by the ratio of the percent absorption of the solution to the percent absorption of the dye if it were present alone. Parker (1957) refers to

---

*The loss in quantum yield may be derived in the following manner: Let q be the quantum yield of emission for an excited molecule, I the quantum intensity of light absorbed initially, and β the fraction of emitted light that is reabsorbed (assume only one reabsorption step occurs). Then qI is the intensity of primary emission and $qI\beta$ is the total reabsorbed intensity. The intensity of primary emission observed is thus $qI(1-\beta)$, to which is added the intensity of secondary emission ($\beta q^2 I$) in order to obtain the total emission observed:

$$F = qI(1-\beta) + \beta q^2 I = Iq(1-\beta+q) \tag{6.12}$$

where F is the total fluorescence intensity observed. Since the average quantum yield ($\bar{q}$) is F/I by definition

$$\bar{q} = F/I = q(1-\beta+\beta q) \tag{6.13}$$

Thus, if q is less than one, the observed quantum yield in concentrated solutions (for which $\beta > 0$) is less than in dilute solution ($\beta = 0$).

this screening effect as an "inner-filter" effect and discusses the phenomenon further.

4. Raman Spectrum of Solvent

This is an emission phenomenon of the pure solvent itself (see Chapter 2). The intensity of this emission is quite small but will interfere in emission spectra any time the instrument sensitivity must be very high (as for a low quantum efficiency of fluorescent dye, or a low concentration of dye). Parker (1959) has shown that the solvents containing hydrogen atoms show a strong Raman band of about 0.3 reciprocal microns (25 to 75 nm) to the red of the exciting light. Table 6.2 gives

TABLE 6.2 SPECTRAL SHIFT OF RAMAN EMISSION BAND IN WATER
SOLVENT[a]

Exciting light, nm	248	313	365	436
Shift, μ^{-1}[b]		0.339	0.339	0.340	0.335
Shift, nm	23	38	52	75
Raman emission				
peak, nm	271	351	417	521

[a]Parker (1959).
[b]Reciprocal microns.

the spectral shifts for water as a function of exciting wavelength as well as the peak wavelength of the strong Raman band. Parker (1959) also gives data for ethanol, cyclohexane, chloroform, and carbon tetrachloride.

5. Photodecomposition

This effect can lead to either a decrease in fluourescence intensity, a distortion of the spectrum, or both. A dye which is unstable under radiation with the exciting light may decompose and give rise to products which absorb at the wavelength of excitation and therefore act as an internal "screen" as discussed above. Most often the photodecomposition product is nonfluorescent. If this degradation product were fluorescent, its fluorescence spectrum would be superimposed upon that of the original dye such that the observed spectrum would change with the time and intensity of excitation. Even without the screening effect the decrease in the amount of fluorescent dye would result in a decrease of fluorescence intensity as a function of the length of time of irradiation.

Photooxidation is a specfic form of decomposition that is dependent upon the presence of oxygen in the solution which reacts with the excited dye molecule to form products which are usually nonfluorescent. This is not to be confused with oxygen quenching of fluorescence in which the dye structure is not altered. The effects of oxygen in the two processes are discussed by Bowen and Williams (1939).

6. *Bimolecular Complexes*

This is a further extension of the dark complex formation referred to above. In some cases, especially when the dye concentration is high, dye molecules will associate and form dimers (see Sec. 2.5). The absorption bands of the dimer may differ considerably from that of the monomer and the dimer is quite often nonfluorescent. However, there are instances in which the dimer formed is fluorescent with an emission band at longer wavelengths than the monomer emission band.

This association of monomers may be avoided by lowering the concentration of the dye in solution, by changing pH to favor dissociation, or by changing to some other solvent system which favors the dissociated monomers.

7. *Temperature*

The effect of temperature on the fluorescence properties of a compound is complex due to the inherent interrelationship of many solution phenomena with temperature. Most often an increase in temperature results in decreased quantum yield because it strongly effects the viscosity of the solute environment, the kinetic energy of both solvent and solute, and the vibrational energy state of the molecule. The resulting change in viscosity with increasing temperature affects the collisional rate between the fluorescent molecules and all other molecules present. The lower viscosity results in more collisions between molecules, thus the rate of collisional quenching (thermal relaxation processes) increases. This results in the generally observed fact that fluorescence decreases as the temperature of the solution increases. The viscosity effect is also very important in the measurement of fluorescence polarization (see Chapter 4). Temperature changes can affect polarization not only through the relationship of viscosity to rotational relaxation time, but also through the change in the mean lifetime of fluorescence as a result of a change in collisional quenching. The amount of energy exchanged in collisional quenching is also a function of temperature, which affects the state of the molecule; the greater energy exchange most probably leading to decreased fluorescence. Since the vibrational state of the molecules are a function of temperature (in accordance with the Boltzmann distribution, (see Chapter 1), and as for most molecules the ground state transition is more probable for absorption of light, this increased vibrational energy would result in a change in the observed decrease in fluorescence.

Furthermore, if the system under study is dependent upon intermolecular interactions (complexes), the interactions themselves may be highly temperature dependent and thus require close temperature control to maintain a constant population of species under study. Protein-ligand interactions fall within this category as well as those systems in which a complex quencher is used either analytically or as an unavoidable contaminant. Dimer formation mentioned above may also be affected by the environmental temperature.

Therefore, it is clearly desirable to remove ambient temperature effects from fluorescence measurements; this has been accomplished by a variety of temperature-regulated cell holders.

When low temperatures are to be employed special precautions must be taken to avoid fogging of the system through condensation of moisture out of the air. The viscosity effect upon fluorescence may be studied by using solvents of high viscosity such as glycerol, but care must be used when extrapolating data from one solvent system to another.

C. Instrumental Errors

1. *Geometry*

This very complex factor in fluorescence analysis is instrumental but is interrelated with the concentration of the absorbing species in a solution being studied. Fluorescence is radially dispersed and with the exception of an integrating sphere only a small portion of the emitted light is seen by an instrument. This portion of light is dependent upon the type of optical design of the instrument. The discussion shall be split into the following categories: (a) right angle fluorometers and (b) front face fluorometers.

a. *Right-Angle Fluorometers* (Fig. 6.3). The proportional relationship between amount of light emitted and the amount of light absorbed is a fundamental observation (if there is no concentration dependence of the quantum yield). This holds true, however, only for the total fluorescence from the sample, which is far from being equivalent to that signal measured as fluorescence intensity in most instruments. To better explain the situation, let us refer to Fig. 6.9, which is representative of the typical geometric situation found in an instrument of the right-angle type.

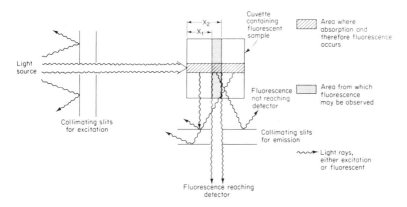

Fig. 6.9. Geometry considerations found in right-angle fluorometer.

It is seen from this representation that even though the fluorescence is not observed from the total absorbing volume it is always measured from the same volume of space represented by the area containing both dots and cross-hatching. This indicates that the observed intensity is related to the total intensity by a proportionality constant and indeed, introduces no error in relative intensities. What is significant, however, is that the relationship between absorption and fluorescence *in that volume* of observation be considered.

In general, this means that as the concentration of a substance increases, a greater and greater proportion of the exciting light is absorbed before ever reaching that volume element from which fluorescence is observed. Offsetting this is the phenomenon that, as the concentration of the substance increases, more of the exciting light reaching that volume is absorbed and more fluorescence is observed. However, it should be obvious that this cannot be true at concentrations approaching infinity but that there must be some concentration above which the fluorescence must decrease because there simply is too little exciting light reaching the observation volume which is not compensated by the greater absorption of that volume. This means that maximum fluorescence (defined as one hundred percent fluorescence) occurs at a concentration which gives less than one hundred percent absorption (shown in Figure 6.10(a)); the exact point at which this occurs will vary between instruments.

This situation may be rendered tolerable in fluorescence assays in several ways. All assays may be run at a constant absorbancy so that geometry changes do not occur. Another way is to simply calibrate the instrument response at increasing concentration with some fluorescent

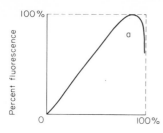 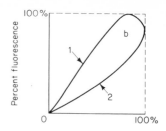

Fig. 6.10. The behavior of fluorometers as a result of the geometry existent in emission optical path and the cuvette, a-right-angle fluorometer as in Fig. 6.9. b-1, geometry as in Fig. 6.11 and b-2, geometry as in Fig. 6.12, both are front surface fluorometers.

standard and therefore comparing a standard to each unknown (both at the same absorbancy). This will constantly correct for changes in geometry due to changes in absorption. It is possible to avoid these correction factors when instrument sensitivity and the quantum yields of the dyes are high. The absorbancy of the solutions may be set at low levels such that the amount of light lost by absorption in the front part of the sample cell is insignificant, and thus for all practical purposes the amount of light reaching the observation volume is constant for all samples. The absorbancies should be below 0.05, as shown in Sec. 6.2B and Eq. (6.5).

Parker and Barnes (1957) give the following method for correcting for this phenomenon to which they refer as the "inner-filter" effect (Sec. 2.4B):

$$F_0 = \frac{F(2.303D)(X_2 - X_1)}{10^{-D_{X_1}} - 10^{-D_{X_2}}} \tag{6.14}$$

in which F is the observed fluorescent intensity, D is the optical density per centimeter path length, and X_2 and X_1 are the distances shown in Fig. 6.9.

b. *Front-Surface Fluorometers.* For the excitation and observation geometry depicted in Fig. 6.11, the relationship between absorption and fluorescence is the same as for a right angle fluorometer and is shown in Fig. 6.10, curve b-1.

Figure 6.12 shows a situation for a front-surface fluorometer which is a better, although still not ideal, system. As is seen, this geometry allows the observation of fluorescence from a volume adjacent to the initial surface of irradiation so that, as the concentration of dye absorbancy increases, the observed fluorescence will increase more rapidly than expected from the increased absorbancy. The rate of increase of fluorescence is greater than a simple linear relationship

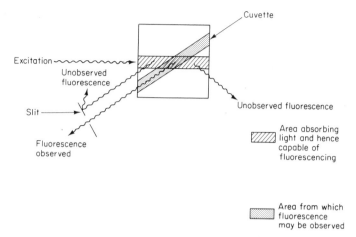

Fig. 6.11. Front-surface fluorometer which has geometry giving rise to behavior identical with a right-angle fluorometer.

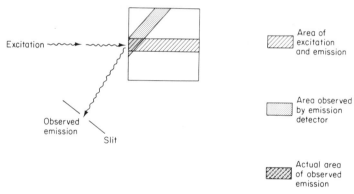

Fig. 6.12. Front-surface fluorometer showing geometry responsible for the response curve shown in Fig. 6.10b2.

because more of the total light absorption occurs in the volume of solution under observation. Effectively, I_0 as well as c in Eq. (6.4) are increasing faster than in the 90° angle system in which, as c increases, I_0 *for the volume observed* decreases due to increased absorption in the preceding solution volume. Figure 6.10 b-2 shows the behavior expected for this type of geometry.

There are several ways in which this geometric effect may be eliminated. Obviously, by making all measurements at a constant light absorbancy, the geometry will be identical for all, and the measured intensities are directly comparable. It is not always practicable to keep the

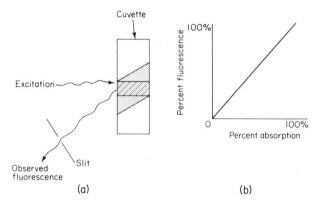

Fig. 6.13. Front-surface fluorometer using a shorter absorption path length so as to give a more ideal fluorescence-absorption response curve: (a) is the arrangement of the cuvette and excitation-fluorescence geometries; (b) shows that this response should be and is identical with the ideal curve.

absorbancy constant, in these instances a characteristic response curve using some standard fluorescent dye for the instrument is obtained so that standards of varying absorbancies can be used as references to correct for instrument response. A simpler method is to keep the optical density so low (< 0.05) that the fluorescence is essentially dependent only on the concentration.

2. *Scattering*

Not all of the exciting light passes directly through a solution but some may be scattered by finely dispersed particles in the solution, by reflection from cuvette surfaces or from the cuvette holder surfaces (if these are not coated with a nonreflective paint) and also by Raleigh scattering of the solvent. The net effect of the scattering process is to direct exciting light into the emission detection optical path. This would give rise to erroneously high intensities of fluorescence if the emission optical system could not eliminate a significant portion of the exciting light. An estimate of the difference in magnitude between the intensity of the excitation and detected light of a sample placed in the usual fluorescence spectrophotometer would be of the order of 10^8. Thus scattering can be a vital factor.

The scattering problem is encountered in all instruments, whether they select the fluorescence to be observed by filters or by a monochromator. There is no simple manner to determine if scattering is taking place and giving abnormally high readings, but it may be checked by the use of a sample solution (a "blank") without the fluorescence due to determine the order of magnitude of the background or scattering signal. A more nearly equivalent measure of the scattering signal is to use a sample solution to which a *nonfluorescent* dye is added such that the absorbancy of the nonfluorescent solution is equivalent to that of the fluorescent solution.

In general, if an instrument is built with a monochromator to separate both the excitation and emission light, and if well-separated narrow bandwidths of emission and excitation are employed, scattering of the excitation light into the fluorescence detector is undetectable. When using an instrument with filters for selecting the fluorescence wavelengths, the greater the separation of the wavelength of excitation from the bandpass region of the filter, the less the likelihood of contamination by the leakage of scattered light through the filters.

Another consideration of filter instruments is that some color filters which absorb light in the ultraviolet will phosphoresce. For example, if the exciting light is ultraviolet and is scattered into the color filter, the luminescence of the filter will add to the fluorescence reaching the detector resulting in too high readings (a secondary effect of the scattering). Therefore, when using ultraviolet excitation and a filter shown to be luminescent under these conditions, an additional filter consisting of a 2- to 4-mm layer of $2 M$ $NaNO_2$ solution is placed between the cuvette and the color filter. This layer will absorb the near ultraviolet and shorter wavelengths but will not affect the visible spectrum.

3. *Polarization of Fluorescence and Optical Systems*

Polarization of light is a phenomenon that occurs when right angle detection systems are used. As elaborated further in Chapter 4, the right angle system observes the $I_\parallel$ and $I_\perp$ components of the emitted light. Since light is emitted in three dimensions the true composition of the fluorescent light when excited with unpolarized light would be $2I_\parallel$ and $1I_\perp$. If the exciting light is polarized this would then be $I_\parallel$ and $2I_\perp$. Since most solutions are not highly polarizing this difference may be ignored. In the case of highly polarizing solutions there may be a significant error in a computation of the total fluorescence emitted from the solution.

In addition, the exciting light reaching the sample cuvette is probably not completely unpolarized because any plane reflecting surface (mirrors, gratings, and prisms) preferentially reflects light polarized parallel to the reflecting surface. Also, lenses used for focusing or collimation may not be completely unpolarized, i.e., they may have greater transmissions for one plane of polarization than for another.

Similarly, the optical system on the emission side may also have a net polarization due to the same factors which obtain for the excitation optics and therefore eliminate that part of the fluorescence which is not of a certain polarization. The net result is that the intensity reaching the emission detector is dependent upon the polarization of the solution. This usually will not make a great difference, but a case where this could be most noticeable would be in the comparative determination of the quantum yield for a dye in solution to that covalently bound to a protein. The polarization of the dye in solution would be close to zero but that bound to the protein could be 0.4.

If such an effect is suspected to be influencing measurements, there is no alternative but to check the polarization of the solution.

6.5 Quantum Yield of Fluorescence

A. DEFINITION AND CHARACTERISTICS

The quantum yield or quantum efficiency of fluorescence is defined as the ratio of quanta emitted to quanta absorbed. It is independent of the energy or wavelength of the quanta and thus differs from the energy efficiency of fluorescence!

$$\text{quantum yield} = q = \frac{\text{quanta emitted}}{\text{quanta absorbed}} \qquad (6.15)$$

In the case of fused ring structures quantum yield is normally independent of the wavelength of the absorbed light since emission occurs from the lowest level of the excited state regardless of the level to which the molecule is raised upon absorption of light (the Franck–Condon excited state — see Sec. 2.3).

Figure 6.14 shows the kinetic dependence of quantum yield. The factors that govern the relative magnitudes of the different rate constants, which may vary markedly from one solvent to another, are not well understood. For example, the dye 1-anilinonaphthalene-8-sulfonic acid in water at pH 7.0 has a quantum yield of about 0.005, but in dimethylformamide, its quantum yield is about 0.85. This indicates that k_f in Fig. 6.14 is much larger than k_k and k_x when the dye is in dimethylformamide than when the dye is in water solution.

B. DETERMINATION OF QUANTUM YIELDS

The quantum yield of a fluorescent dye is determined either by a direct absolute measurement or by comparison to a reference dye. The

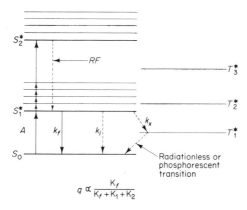

Fig. 6.14. T_1^* = 1st excited triplet; S_0 = singlet ground state; S_1^* = 1st excited singlet state; S_2^* = 2nd excited singlet state; A = possible absorption transitions (Franck–Condon); RF = radiationless energy loss to lowest excited singlet from which emission occurs; k_f = probability of radiative energy loss in $S^* \rightarrow S_0$ transition; k_i = probability of radiationless $S^* \rightarrow S_0$ transition; k_x = probability of transition from S^* to nonfluorescent T^* state.

direct method is not usually the more desirable but is subject to greater experimental error and is more tedious. The comparative approach is normally dependent upon other work in which a direct measurement has been made. It is most common to use a dye whose quantum yield and spectrum are well established and which is easily purified. An example of such a dye is quinine sulfate.

1. *Absolute Quantum Yields*

a. *Integrating Sphere.* In this method all of the fluorescence light from a sample is collected by exciting the sample within a spherical chamber that is coated with some highly reflective and dispersive substance. This coating is usually magnesium oxide or barium sulfate. Figure 6.15 shows the integrating sphere arrangement used by Förster and Livingston (1952).

It is important to notice that Förster and Livingston use a thermopile as a detector. This instrument combination necessitates a correction from energy to quanta units but has the advantage of a somewhat broader range of sensitivity. It is possible to substitute a photomultiplier for the thermopile for better sensitivity in the ultraviolet. However, this results in less sensitivity in the visible spectrum. The following is a summary of the calculations involved to extract quantum yields from an integrating sphere, when the detector used is a thermopile. In

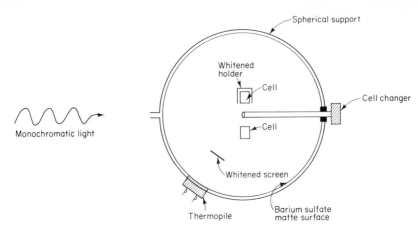

Fig. 6.15. Integrating sphere after Förster and Livingston (1952).

general, the light intensity reflected from a given area of the wall is given by

$$E = cI/\alpha \qquad (6.16)$$

where E = energy striking a given area of the wall
         I = intensity of source
         α = absorption coefficient of the wall
         c' = constant of proportionality

The source is considered to be the cuvette upon which the exciting light falls and from which fluorescence is emitted. To measure the intensity of the exciting light absorbed, a filter is placed before the detector which passes exciting light but not fluorescent light. The difference in detector signals between the blank cuvette and the sample cuvette is proportional to the intensity of light absorbed:

$$I_A = \frac{\Delta A \alpha_E}{c' G_E} = \text{intensity of light absorbed} \qquad (6.17)$$

where ΔA = difference in detector response between solvent cuvette
             and sample cuvette
         α_E = absorption coefficient of wall for exciting light (λ_E)
         G_E = transmission coefficient of filter for exciting light (λ_E)
         c' = constant of proportionality

The same procedure is used to measure the intensity of fluorescence except that a filter is placed before the detector which absorbs exciting light but transmits the fluorescent light. The fluorescent light intensity is proportional to the difference in detector response when the solvent cuvette and when the sample cuvette are placed in line with the exciting light:

$$I_F = \frac{\Delta F \alpha_F}{c' G_F} = \text{fluorescent light intensity} \qquad (6.18)$$

where ΔF = difference in detector response
       α_F = absorption coefficient of wall for α_F
       G_F = transmission coefficient of filter for α_F
       c' = constant of proportionality

G_F and α_F must be integrated coefficients if only a filter is used before the detector, since emission is over a range of wavelengths. If a mono-chromator-photomultiplier arrangement (spectrometer) is used, I_F would have to be an integrated function of the wavelengths of fluores-cence. For a spectrometer,

$$I_F = \frac{1}{c'} \frac{\Delta F}{G_F} \frac{\alpha_F}{G_F} = \frac{1}{c'} \sum_{\lambda_1}^{\lambda_n} \frac{\Delta F(\lambda)\alpha_F}{G_F} d\lambda \qquad (6.19)$$

where λ is the wavelength of fluorescence as selected by the spectro-meter, the sum being taken from the shortest wavelength λ_1 at which fluorescence is detectable to the largest wavelength λ_n of detectable fluorescence at wavelength intervals $d\lambda$.

The quantum yield of fluorescence may now be calculated:

$$\text{energy efficiency of fluorescence} = \frac{I_F}{I_A} \qquad (6.20A)$$

$$\text{quantum yield of fluorescence} = q = \frac{I_F \lambda_F}{I_A \lambda_E} \qquad (6.20B)$$

where λ_F = wavelength of fluorescence
       λ_E = wavelength of excitation

Substituting the expressions of Eqs. (6.18) and (6.19) into (6.20B):

$$q = \underbrace{\frac{\Delta F \alpha_F \lambda_F G_E}{G_F \Delta A \alpha_E \lambda_E}}_{\substack{\text{when using} \\ \text{filters for} \\ \text{fluorescence} \\ I_F \text{ as in Eq.} \\ (6.18)}} = \underbrace{\frac{G_E}{\Delta A \alpha_E \lambda_E} \sum_{\lambda_1}^{\lambda_n} \frac{\Delta F(\lambda)\alpha_F(\lambda)\lambda_F \, d\lambda}{G_F(\lambda)}}_{\text{when spectrometer is used}} \qquad (6.21)$$

The calculation of q will vary somewhat, as has been implied, if a spectrometer is used, but references to the previous paragraph and to the discussion in Sec. 6.3 will describe how to allow for the necessary recalculations.

b. *Fluorescent Screen.*   This method has seemingly been favored over the integrating sphere method because the instrumentation is more readily available and easier to construct. Figure 6.16 shows one

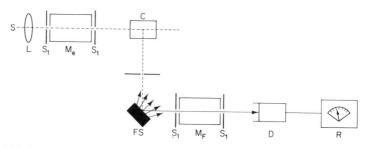

Fig. 6.16. Instrumentation for measurement of quantum yield by use of fluorescent screen: S = light source; S_1 = adjustable slits; L = focusing lens; C = sample cuvette; FS = fluorescent screen; D = detector (thermopile, photocell, or photomultiplier tube); R = recorder or galvanometer; M_e = excitation monochromator or filters; M_F = emission monochromator or filters to pass fluorescence light from screen.

of many possible arrangements for measuring quantum yields by this method. The fluorescent screen is a sample cell containing a fluorescent dye with a broad absorption spectrum. The concentration of this dye must be high enough to ensure complete absorption of the exciting light. If this requirement is met, the fluorescence intensity from the fluorescent screen is linearly dependent upon the intensity of light striking it from the sample solution since all incident light is absorbed. The use of a blank solution in the sample cell will allow one to correct approximately for any signal due to exciting light scattered from the front surface.

There are two ways to measure the intensity of the exciting light. One is to use a MgO screen to scatter the incident radiation onto the fluorescent screen; this method has been discussed by Bowen and Sawtell (1937) and Bowen and Williams (1939). However, Gilmore et al. (1952) point out the difficulty in calculating the difference in the spatial distribution between the light from the diffusing screen (MgO) and the fluorescent solution. For this reason, Weber and Teale (1957) used a scattering solution containing glycogen. With this method the geometry is more similar to that found in the fluorescent solution since the dipole scattering due to the glycogen attenuates the beam of exciting light in the same manner as an absorbing medium.

The method used by Weber and Teale (1957) requires that the polarization of the scattering solution and of the sample solution must be measured. This measurement is not necessary with the use of a scattering screen.

Since the calculations are fundamentally the same in this case as for the integrating sphere, they will not be repeated here. They may be obtained from the above references. Gilmore et al. (1952) obtained their results in a mixed solvent glass of ether, isopentane, and ethanol (E.P.A.) at 77°K and, in addition, had to make refractive index corrections since emission was observed at 45° to the face of the cuvette. Melhuish (1955) has determined the quantum yield of quinine bisulfate, an often used fluorescence standard.

2. *Relative Quantum Yields*

This method is by far the simplest method for obtaining quantum yields. It consists simply of comparing the fluorescence intensity of the sample under study to the intensity of a dye of known quantum yield. If a spectrometer (monochromator-photomultiplier) is used to detect the fluorescence, the measurement is already in terms of quanta and no corrections need be made for converting energy to quanta. Equations (6.22) and (6.23) allow us to see the restrictions on the method:

$$F = I_E q \%A G(\theta) \qquad (6.22)$$

where F = observed fluorescence of dye solution
$G(\theta)$ = geometry factor (< 1) since not all of the fluorescent light is observed
I_E = intensity of exciting light
q = quantum yield
$\%A$ = percent absorption of solution $(100 - \%T)$

The ratio of the quantum yield of the standard dye (q_s) and the unknown (q_x) is

$$\frac{F_x}{F_s} = \frac{q_x}{q_s} \qquad (6.23)$$

where F_x is the measured fluorescence of the unknown and F_s is that of the standard. In order for (6.23) to be true, the quantity

$$\frac{I_{Ex} \%A_x G(\theta)_x}{I_{Es} \%A_s G(\theta)_s}$$

must be equal to unity. The best way to ensure this is to excite both standard and sample at the same wavelength and to have the solutions of equal absorbancy at this wavelength; we then have

$$\frac{I_{Ex}\%A_x}{I_{Es}\%A_s} = 1$$

The ratio of geometry factors will also be unity $[q(\theta)$ of the unknown $= q(\theta)$ of the standard] since the same instrument is used.

Finally, the fluorescence ratio in (6.23) must be discussed. If a fluorescent screen detector system is used, no corrections need be made for the instrument response since the response is directly proportional to the total or integrated fluorescence intensity spectrum. However, if a spectrometer system is used, one must integrate the total area under the *corrected* emission spectrum for both sample and standard and use the ratio of these areas in (6.23). In practice, this integration is oftentimes unnecessary if it can be demonstrated that the ratio of observed peak fluorescence intensities (corrected for spectrometer efficiency) is the same as the ratio of the integrated corrected fluorescence intensities, thus simplifying the method even more.

There are many possible dyes from which to choose reference standards for quantum yield determinations. As mentioned before the most commonly used is quinine bisulfate in 0.1 to 1 M sulfuric acid because the dye is easily purified and recrystallized, will not associate in acid, and is very stable. Its quantum yield, as determined by Melhuish (1955), is 0.55.

Table 6.3 is a list of some of the quantum yields reported for various dyes. Most values are derived from calibrated instruments but several presented have been obtained by the relative method.

6.6 Fluorescence Excitation Spectra

A. Purpose

The fluorescence excitation spectrum of a substance is obtained by measuring the intensity of fluorescence as a function of the wavelength of excitation. Since fluorescence is directly proportioned to $I\epsilon q$, then scanning a dye solution at low concentration with the excitation monochromator while holding the fluorescent monochromator at the wavelength of maximum fluorescence will give a spectrum that is directly proportional to ϵ, the absorption coefficient. This method is a very

TABLE 6.3 QUANTUM YIELDS OF SOME IMPORTANT COMPOUNDS

Compound	Solvent	q	References
Benzene ($0.03M$)	E.P.A.[a] (77°K)	0.24 ± 0.03	b
Naphthaline	Alcohol	0.12	c
Naphthaline ($10^{-3}M$)	E.P.A. (77°K)	0.47 ± 0.03	b
Fluorescein		0.84	d
Fluorescein ($10^{-5}M$)	ag. NaOH	0.79 ± 0.06	e
		0.92	c
Fluorescein	ag. NaOH	0.85	f
		0.85	g
Eosin ($10^{-5}M$)	Water	0.12 ± 0.02	c
		0.23	g
		0.16	d
		0.19	c
Magdala red ($10^{-5}M$)	Ethanol	0.49 ± 0.04	e
Magdala red ($10^{-5}M$)	Ethanol	0.56	d
Rubrene ($10^{-5}M$)	n-heptane	1.02 ± 0.08	e
Rubrene ($10^{-5}M$)	n-heptane	1.0	d
Rhodamine B	Ethanol	0.69	g
Rhodamine B	Ethanol	0.89 (excitation at 535 mμ)	c
Rhodamine B	Ethanol	0.97 (excitation at 366 mμ)	c
Chlorophyll A	Benzene	0.325	c
		0.26	c
Chlorophyll B	Benzene	0.11	c
Riboflavin	Water, pH7	0.26	c
		0.06	h
Riboflavin	Water pH7	0.26	c
Fluorene	Ethanol	0.53	c
N-methyl acridinuum chloride	Water	1.01	c
Acriflavin	Water	0.54	c
9-amino-acridine	Water	0.98	c
Shenal	Water	0.22	c
Indole	Water	0.45	c
Skatole	Water	0.42	c
Pyrene	Poly (methyl methacrylate)	0.61	i
Sodium ealicylate	Water	0.28	c
Phenanthrene	Alcohol	0.10	c
Anthracene	Ethanol	0.28	
Anthracene	Benzene	0.29	c
Anthracene	Poly (methyl methacrylate)	0.24	j
Anthracene	Benzene	0.24	j
Fluorobenzene	E.P.A. (77°K)	0.25	b
Chlorobenzene	E.P.A. (77°K)	0.08	b

TABLE 6.3 (*Contd.*)

Compound	Solvent	q	References
Proflavin	H_2O pH4	0.27	*j*
2 Naphthylamine	Benzene	0.486	*k*
2-Naphthol	Water, pH10	0.21	*j*
1-Dimethylamino-naphalene 4 sulfonate	Water	0.48	*c*
Quinine bisulfate	1.0 *N* sulfonic acid	0.55	*j*
Quinine bisulfate	1.0 *N* sulfonic acid	0.51	*k*

[a] Ether, isopentane ethanol solvent.
[b] Gilmore et al. (1952, 1955).
[c] Weber and Teale (1957).
[d] Vavilov (1924).
[e] Förster and Livingston (1952).
[f] Umberger and LaMer (1945).
[g] Parker and Rees (1960).
[h] Ellis and Rogers (1962, 1964).
[i] Strickler et al. (1956).
[j] Melhuish (1964).
[k] Melhuish (1961).

sensitive means of measuring the absorption spectrum of a substance, provided of course that it is fluorescent.

The advantage of this method over that of a pure absorption spectrum is twofold. Whereas absorption measurements are good only to a concentration of about $10^{-5} M$, fluorescence excitation spectra may be made at concentrations as low as $10^{-8} M$, with good instrument sensitivity and low noise-to-signal ratio. Secondly, the absorption spectrum of a fluorescent compound which is in solution with another substance that absorbs in the same region, but is not fluorescent, may be measured. This latter fact allows an investigator to obtain the concentration of a dye under the circumstances just iterated, simply by calibrating instrument response to the concentration of the dye, since response will be proportional only to the absorption coefficient, and, hence, the concentration (or optical density) of the dye, not to the total optical density of the solution.

For further discussion of fluorescence excitation spectra see the general references and Weber and Teale (1958) and Parker and Rees (1960).

B. Measurement of Excitation Spectra

The fluorescence intensity of a solution is given by Eq. (6.5). The two quantities in this expression which are dependent upon the wavelength of excitation are ϵ_λ (which is the quantity we are seeking to measure) and I_λ. The latter will vary with wavelength because the lamp does not have equal quantum output at all wavelengths, nor is the efficiency of the excitation monochromator constant over all wavelengths.

Section 6.3A discusses the method for obtaining the combined correction curve for lamp output and monochromator efficiency. Figure 6.17 is a comparison of absorption and excitation spectra of

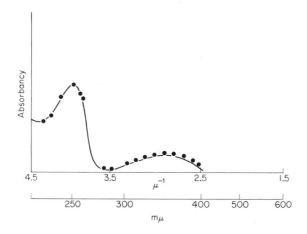

Fig. 6.17. Absorption spectrum (full curve) and the true excitation spectrum (circles) of quinine bisulfite in 0.1 *N* sulfuric acid (Parker and Rees, 1960).

quinine bisulfate, as published by Parker and Rees (1960). The excitation spectrum is corrected by the curve shown in Fig. 6.5(a).

The procedure by which an experimentor obtains the necessary corrections for the lamp-monochromator output using a standard lamp may become less accurate as the lamp filament ages. Also, the actual application of these corrections to the observed spectrum is both laborious and time consuming, making an easier method very desirable. An easy, efficient means to obtain automatic correction for variations in lamp-monochromator output is to feed into one arm of a ratio recorder the signal from a photomultiplier that continuously monitors the light intensity of the beam entering the sample from the excitation monochromator; into the other arm of the ratio recorder the signal from the fluorescence photomultiplier is fed. The spectra thus directly obtained have no need to be corrected and are true absorption spectra. Parker (1958) and Weber and Young (1964) have described instruments of this type.

6.7 Determination of the Number of Components in a Fluorescent Solution

When using fluorescence spectroscopy it is important to know if the observed fluorescence arises from a simple or a mixture of emitting species, which are behaving independently of each other in the absorption and emission processes. The following discussion describes a method of determining the number of emitting species in a fluorescent solution. A more complete description of the mathematics involved may be found in any good matrix algebra text [such as McLachlan (1963)] some of which are listed at the end of the chapter. Use of this method is also described in a paper by Weber (1961).

Let us assume that the intensity of fluorescence observed on exciting at wavelength i and recorded at wavelength j is dependent upon both these conditions, $F(i, j)$. Furthermore, F is related to the sum of the fluorescence intensities of each of k dye components present in the system:

$$F(i, j) = \sum_k F_{i,j}^k \tag{6.24}$$

$F_{i,j}^k$ is the fluorescence of the kth component of the solution which is excited at wavelength i and emitting at wavelength j. The superscript k identifies the component whereas subscripts identify the wavelengths (Fig. 6.18).

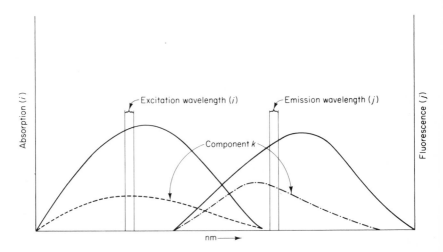

Fig. 6.18. Absorption and emission for component k and for entire solution. Excitation at i, emission recorded at j.

$F^k(i,j)$ is considered to depend upon two variables, one a function of exciting wavelength only (f_i^k) and one of emission wavelength only (q_j^k):

$$F^k(i,j) = f_i^k q_j^k \qquad (6.25)$$

f_i^k includes the fraction of exciting light absorbed by component k. q_i^k involves mainly the quantum efficiency of k as well as the fraction of total intensity emitted at the one wavelength j. Any geometric factors are included in the two functions as constants and do not affect further discussion. Combining (6.24) and (6.25) results in the following definition of $F(i,j)$

$$F(i,j) = \sum_k F^k(i,j) = \sum_k f^k(i)q^k(i) \qquad (6.26)$$

Since $F(i,j)$ is now defined for any combination of wavelengths it is possible to analyze what happens to $F(i,j)$ when i and j are varied. If F is measured at m wavelengths of excitation and n wavelengths of emission, the following matrix or array is obtained;

$$|F|_{mn} = \begin{vmatrix} F_{11} & F_{12} & F_{1j} & F_{1n} \\ F_{21} & & & \\ F_{i1} & & F_{ij} & \\ F_{m1} & & & F_{mn} \end{vmatrix} \qquad (6.27)$$

Into each element of (6.27) expression (6.26) may, of course, be substituted. Choosing a particular 2×2 minor of (6.27) and replacing $F(i,j)$ according to (6.26):

$$|F|_{22} = \begin{vmatrix} F_{11} & F_{12} \\ F_{21} & F_{22} \end{vmatrix} = \begin{vmatrix} \sum f_1^k q_1^1 & \sum f_1^k q_2^k \\ \sum f_2^k q_1^k & \sum f_2^k q_2^k \end{vmatrix} \qquad (6.28)$$

The value of the above determinant (henceforth, a square matrix will be called a determinant) is

$$D = \sum_k f_1^k q_1^k \sum_k f_2^k q_2^k - \sum_k f_1^k q_2^k \sum_k f_2^k q_1^k \qquad (6.29)$$

The value of D above will be zero or nonzero depending upon the number of components present. If $k = 1$, then (6.29) reduced to (6.30):

$$\underset{k=1}{D} = f_1^1 q_1^1 f_2^1 q_2^1 - f_1^1 q_2^1 f_2^1 q_1^1 = 0 \qquad (6.30)$$

If $k = 2$, (6.29) reduces to (6.31):

$$\underset{k=2}{D} = (f_1^1 q_1^1 + f_1^2 q_1^2)(f_2^1 q_2^1 + f_2^1 q_2^2) - (f_1^1 q_2^1 + f_1^2 q_2^2)(f_2^1 q_1^1 + f_2^2 q_1^2) \qquad (6.31)$$

Since $f_i^1 \neq f_i^2$ and $q_j^1 \neq q_j^2$, D of (6.31) above will not be zero. The same treatment of the above for $k > 2$ will show that for all $k > 1$, (6.29) is nonzero.

Similarly, taking a 3×3 minor of (6.27), as in (6.32):

$$|F|_{33} = \begin{vmatrix} \Sigma\, f_1^k q_1^k & \Sigma\, f_1^k q_2^k & \Sigma\, f_1^k q_3^k \\ \Sigma\, f_2^k q_1^k & \Sigma\, f_2^k q_2^k & \Sigma\, f_2^k q_3^k \\ \Sigma\, f_3^k q_1^k & \Sigma\, f_3^k q_2^k & \Sigma\, f_3^k q_3^k \end{vmatrix} \qquad (6.32)$$

expanding by minors to evaluate (6.32) yields (6.33):

$$D_{3\times3} = \Sigma\, f_1^k q_1^k \begin{vmatrix} \Sigma\, f_2 q_2 & \Sigma\, f_2 q_3 \\ \Sigma\, f_3^k q_2^k & \Sigma\, f_3^k q_3^k \end{vmatrix} - \Sigma\, f_1^k q_2^k \begin{vmatrix} \Sigma\, f_2^k q_1^k & \Sigma\, f_2^k q_3^k \\ \Sigma\, f_3^k q_1^k & \Sigma\, f_3^k q_3^k \end{vmatrix} \qquad (6.33)$$

$$+ \Sigma\, f_1^k q_3^k \begin{vmatrix} \Sigma\, f_2^k q_1^k & \Sigma\, f_2^k q_2^k \\ \Sigma\, f_3^k q_1^k & \Sigma\, f_3^k q_2^k \end{vmatrix}$$

In the above expression (6.33), if k is 1 or 2, D will be zero, if $k \geqslant 3$, D is nonzero.

In general, the determinant of lowest rank yielding a zero value has a rank one greater than the number of components. More explicitly, if

$$D_{n-2 \times n-2} \neq 0, \qquad D_{n-1 \times n-1} \neq 0, \qquad D_{n \times n} = 0, \qquad D_{n+1 \times n+1} = 0$$

then

$$k = n - 1$$

The problem still remains to decide when the value of a determinant is zero since there is experimental uncertainty involved in all these measurements. The following criteria have been suggested by Weber (1961). Let e be the average error in determining the average value of the matrix element $\bar{F}$. In a 2×2 determinant, of value 0, the calculated value will differ from zero as follows:

$$D = 0 \pm 2\bar{F}e \qquad (6.34)$$

Furthermore, if one defines a parameter P as being the *sum* of the determinant products, rather than the difference, then the ratio of D to P allows estimation of departure from zero ($P \sim 2\bar{F}^2$):

$$\frac{D}{P} = \frac{\pm 2\bar{F}e}{2\bar{F}^2} \sim \pm\frac{e}{\bar{F}} \qquad (6.35)$$

It is suggested that the ratio in (6.35) if less than three times the experimental variation, the determinant may be considered to have a value of zero.

At this point, it is desirable to refer back to Eq. (6.31). Let us now consider two cases in which two-component systems yield zero 2×2 determinants.

For case 1, we allow two different absorbing species to be present, only one of which is fluorescent ($q^1 = 0$, $q^2 > 0$). Equation (6.31) will therefore reduce to (6.30) which is zero. We therefore see that the analysis yields only the number of fluorescent species. The same result applies if we let $q^1 = q^2$ rather than setting $q^1 = 0$.

In case 2, we allow only one component to absorb light but assume it can undergo some excited state dissociation yielding two fluorescent species (such as occurs in naphthol, naphthylamine, etc, see Sec. 1, this chapter). When $q^1 \neq q^2 > 0$ and $f^1 = 0$, $f^2 > 0$ or $f^1 = f^2$ are substituted into (6.31), a value of zero results. Thus the analysis will show that only one ground state species exists even though the fluorescence spectrum indicates the presence of two fluorescing species, the dichotomy of species occurring only after excitation.

6.8 Determination of the Lifetime of the Excited State

A. PHYSICAL PRINCIPLES

The subject of the lifetime of fluorescence has been mentioned previously (Chapter 2). This quantity is an essential value in a discussion of quenching of fluorescence and polarization of fluorescence (see Chapter 4).

Studies on the lifetime of the excited state can give significant information about the kinetics of intermolecular reactions such as dimer and excimer formation (Birks and Monroe, 1967), energy transfer and molecular distances (Stryer, 1968), intermolecular rotational diffusion (Brownian motion), energy transfer and intermolecular distances (Spencer and Weber, 1970).

In order to define the lifetime of fluorescence, let us consider the fluorescent species to be excited by a pulse of light infinitely narrow in time. The relaxation of the fluorescent species from the excited to the ground state follows the exponential decay law:

$$I_t = I_0 e^{-kt} \tag{6.36}$$

where I_t and I_0 are, respectively, the intensities at time t and 0, k

is the rate constant of decay, and t is the time at which I_t is measured. Let us now consider the point in time after excitation at which $I_t = (1/e)I_0$. Substituting this value of I_t into (6.36):

$$\frac{1}{e}I_0 = I_o e^{-kt}$$

$$\frac{1}{e} = e^{-1} = e^{-kt} \qquad (6.37)$$

$$1 = kt$$

$$k = \frac{1}{t}$$

We define this time t at which I_t has reached $1/e$ of I_0 to be the fluorescence lifetime τ. Therefore k is the reciprocal of τ, so that Eq. (6.36) becomes

$$I_t = I_0 e^{-t/\tau} \qquad (6.38)$$

A brief description of spectroscopic measurements of the lifetime of the excited state as originally formulated by Einstein (1917), and modified by other workers (Perrin, 1926; Lewis and Kasha, 1945; Förster, 1951; Strickler and Berg, 1962; Birks and Dyson, 1963), is given in Chapter 2.

B. MEASUREMENT OF LIFETIMES

1. *Experimental Decay*

This method employs Eq. (6.38). A light pulse of short duration (D'Allesio et al., 1964; Hundley et al., 1967) is used to excite a solution containing a fluorescent compound. The emitted light is then monitored by a photomultiplier whose signal is displayed as a function of time on a fast oscilloscope, the trace being recorded photographically. If the light pulse for excitation is assumed infinitely short in duration, a plot of $\ln[I_0/I_t]$ versus t will give a line of slope $1/\tau$ (Fig. 6.19). If more than one mode of fluorescence decay is present, and if the lifetimes of each mode are reasonably disparate, a break in the line will occur such that each mode gives a straight line of slope $1/\tau$. A curved line indicates either nonexponential decay or two more modes of decay not sufficiently different in lifetime to be resolved. This method is therefore quite good for determining heterogeneity within the fluorescent population.

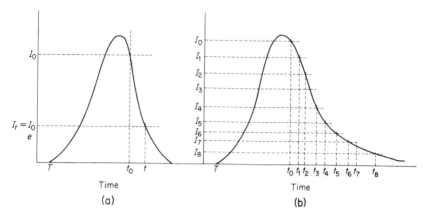

Fig. 6.19A. (a) A representation of a fast oscilloscope tracing of the intensity of light from a pulse light source. $t - t_0 = \tau_{\text{pulse}}$. A plot to give τ from the above trace is shown in Fig. 6.19B(a). (b) A representation of a fast oscilloscope tracing of the intensity of emitted light from a fluorescent solution containing two species of differing lifetime and excited by a pulse of characteristic similar to (a) above. (b) of Fig. 6.19B shows the plot to yield the lifetimes of the components.

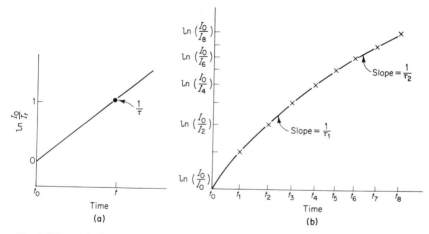

Fig. 6.19B. (a) A plot to yield the decay time of the pulse of light in Fig. 6.19A(a). (b) A plot to yield the decay time(s) of the fluorescent solution giving the decay curve represented in Fig. 6.19B(b).

There are however, serious drawbacks to this method which necessitates caution. For example, one cannot assume infinitely brief pulses of exciting light. In fact, the exciting pulse may, in practice, have a decay time of 2 to 5×10^{-9} sec (see Fig. 6.19). This lifetime may be measured by using a scattering solution to observe the exciting light pulse and determining τ_{pulse} the same way as for τ above. τ must now be corrected for τ_{pulse} to yield τ_{fluor}:

$$\tau \sim \tau_{fluor} + \tau_{pulse} \qquad (6.39)$$

The method for calculation for τ_{fluor} is to use a convolution integral. At this time, measurement of lifetimes of the order of 1 nsec and smaller is very difficult if not altogether impossible. Not only are the pulse lifetimes of this order, but also the response time of most photomultipliers available are in this range. This photomultiplier response time factor puts a lower limit of about 1 nsec on lifetimes measured by this technique (Birks and Monroe, 1967). This limit applies even when the best photomultiplier tubes available are used. For these short times it is necessary to resort to computer averaging to separate signal and noise, to obtain separation of τ_{pulse} and τ_{fluor}. Table 6.4 gives values of lifetimes of several compounds obtained by this method. However, according to Spencer and Weber (1969) using modulation techniques lifetimes on the order of 0.3 nsec can be accurately measured.

TABLE 6.4 Some Fluorescence Decay Times Measured by the Pulse Method

Compound	Solvent	τ (nsec)	Reference
Acridone	100% ethanol	12.48	a
Acridone	95% ethanol	13.0	a
Acridone	Water	15.2	a
Acridone, saturated solution	0.01 M Tris-Cl⁻, pH 7.0	15.4	b
Fluorescein	0.1 N NaOH	4.62	a
Fluorescein ($10^{-6} M$)	0.01 N NaOH	4.5	b
NADH	0.1 N NaHCO₃	4.5	b
Quinine bisulfate	1 N H₂SO₄	15.2	a
Quinine ($10^{-5} M$)	0.1 N H₂SO₄	19.0	b

[a]Ware and Baldwin (1964).
[b]Chem et al. (1967).

2. Modulation of Light

In the above pulse technique, the fluorescent species is excited with pulses at time intervals much greater than τ, so that all the molecules return to the ground state before they are excited again. In this case ω, the frequency of excitation, is much less than k, the rate constant in Eq. (6.36). In this situation, the response is unattenuated and phase is preserved (Fig. 6.20). That is, the fluorescence response follows the

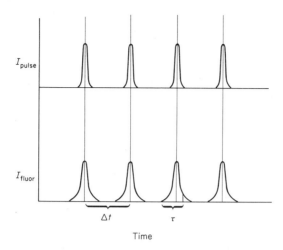

Fig. 6.20. The case in which the fluorescent decay is unattenuated and in phase with the exciting light. $1/t = \omega$, $1/\tau = k$; therefore $\omega < k$.

intensity of the exciting pulse; there is no superpositionary of fluorescence from interfering pulses. However, for ω approaching k and for $\omega > k$ using sinusoidally modulated light, the exponential detector (the fluorescent species) progressively lags behind the exciting light in phase and becomes more attenuated (Fig. 6.21). At $\omega/k = 1$, the phase of the emitted light lags behind the excitation light by 45°. The lag

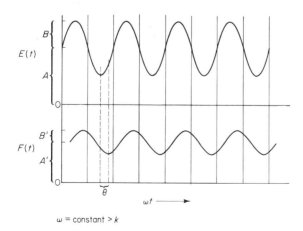

Fig. 6.21. The relationship between frequency of excitation and frequency of fluorescence. $E(t)$ and $F(t)$ are defined as in Eqs. (6.40) and (6.41). θ is the phase shift or lag between excitation and emission $(A' - B')/(A + B)$ = modulation of fluorescence. $(A - B)/(A + B)$ = modulation of excitation [Eq. (6.43)]. ω = constant $> k$.

approaches 90° as attenuation tends toward completion. [See Dushin-sky (1933) for a Fourier spectrum characteristic of an exponential detector].

The theory and further discussion behind instruments of this type may be found in the References (Gaviola, 1927; Maerks, 1938; Bauer and Rozwadowski, 1959; Muller et al., 1965, Spencer and Weber, 1969; Birks and Dyson, 1961; Birks and Monroe, 1967).

If the excitation of a fluorescent species with modulated light of frequency f is described by the expression

$$E(t) = A + B \cos 2\pi ft \tag{6.40}$$

where $A > B$, the modulation of the exciting light is B/A. The fluorescence, if due to exponential decay $(e^{-t/\tau})$ of the population, may be described by the expression

$$F(t) = A + B \cos \theta \cos(2\pi ft - \theta) \tag{6.41}$$

and

$$\tan \theta = 2\pi ft = \omega\tau \tag{6.42}$$

$$\text{relative modulation} = M = \frac{\text{modulation of fluorescence}}{\text{modulation of excitation}}$$

$$= \cos \theta = (1 + 4\pi^2 f^2 \tau^2)^{-1/2}$$

$$= (1 + \omega^2 \tau^2)^{-1/2} \tag{6.43}$$

One may thus measure τ by two methods, measurement of θ or of M (Fig. 6.21). The τ obtained should be the same in both cases if only one fluorescent species is present. The presence of two or more species is indicated by divergence in the values of τ measured by the two parameters, the lifetime determined by modulation being longer than the weighted average of the lifetimes, while that determined by phase being shorter than the weighted average (Spencer and Weber, 1969).

Table 6.5 includes some lifetimes measured by the phase shift method.

C. POLARIZATION

This method of determining the fluorescence lifetime is indirect and involves measuring the polarization of fluorescence as well as the limiting polarization of fluorescence (see Chapter 4 for a discussion of fluorescence polarization). In addition, the rotational relaxation time of

TABLE 6.5 SOME FLUORESCENCE LIFETIMES MEASURED BY THE PHASE SHIFT
METHOD

Compound	Concentration	Solvent	τ(nsec)	Reference
Acridone	—	95% EtOH	13.1	a
Acridone	1% saturation	H_2O	15.6	b
Flavin-adenine dinucleotide	—	0.1 M phosphate, pH 7	2.3	c
Flavin monunucleotide	—	0.1 M phosphate, pH 7	4.65	c
Fluorescein	$10^{-6}\,M$	0.01 M KOH	3.8	b
Fluorescein	10 μg/ml	0.01 M	4.7	c
	1 μg/ml	0.01 M NaOH	4.2	c
NADH	—	0.1 M phosphate, pH 7.5	0.38	c
Quinine	—	$N\ H_2SO_4$	20.1	a

[a]Birks and Dyson (1963).
[b]Muller et al.(1965).
[c]Spencer and Weber (1969).

the fluor must be known (or calculable). The following will demonstrate this method.

Perrin (1926, 1929) developed the following equation which is discussed in Chapter 4:

$$\left(\frac{1}{p}\pm\frac{1}{3}\right) = \left(\frac{1}{p_0}\pm\frac{1}{3}\right)\left(1+\frac{3\tau_0}{\rho_0}\right) \tag{6.44}$$

where p is the polarization measured in the presence of rotational motion, p_0 the limiting polarization in the absence of rotational motion, τ_0 is the lifetime, and ρ_0 the rotational relaxation time. When ρ_0 has been obtained (see Chapter 4) by extrapolation of p to infinite viscosity, the quantity τ_0/ρ_0 may be obtained.

Assuming, for simplicity, the fluor to be a sphere, we have

$$\rho = \frac{f_r}{2kt} \tag{6.45}$$

where f_r is the rotational frictional coefficient, which, for a sphere, according to Kirchhoff, is

$$f_r = 8\pi\eta r^3 \tag{6.46}$$

where η is the viscosity coefficient and r is the radius of the sphere. Thus, (6.45) becomes, after substitution of (6.46),

$$\rho = \frac{4\pi\eta r^3}{kT} = \frac{4\pi N\eta r^3}{RT} \tag{6.47}$$

where N is 6.02×10^{23}, and

$$\left(\frac{\tau_0}{\rho_0}\right) = \tau_0 \left(\frac{1}{\rho}\right) \qquad \text{(6.48a)}$$

$$\tau_0 = \rho \left(\frac{\tau_0}{\rho_0}\right) = \frac{4\pi N \eta r^3}{RT} \left(\frac{\tau_0}{\rho_0}\right) \qquad \text{(6.48b)}$$

If v is the molecular volume, as determined from viscosity measurements, and the molecule is assumed to be a sphere,

$$v = \tfrac{4}{3} \pi r^3 \qquad \text{(6.49a)}$$

and hence by substitution into (6.48b)

$$\tau_0 = \frac{Nv}{2RT} \left(\frac{\tau_0}{\rho_0}\right) \qquad \text{(6.49b)}$$

$$\tau_0 = \frac{V}{3RT} \left(\frac{\tau_0}{\rho_0}\right) \qquad \text{(6.49c)}$$

where V is the molar volume, Nv.

This method is subject to the uncertainties inherent in the measurement of molecular volume as well as the approximation that the fluorescent molecules are spherical. This method does not allow a clear indication of the heterogeneity of lifetimes in the emitting population. Table 6.6 compares the lifetimes calculated by Perrin (1929) with lifetimes measured by the phase shift methods.

D. Possible Sources of Errors

In addition to the errors and inaccuracies mentioned in the previous sections, there are other potential sources of error external to the technique of the measurement itself.

1. Concentration Dependence of τ

It has been noticed in the case of certain fluorescent dyes that the lifetime of fluorescence lengthens with an increase in concentration. It is known from studies of the absorption, excitation, and emission spectra that the phenomenon is not due to formation of dimers or to the presence of excimers. In addition, this phenomenon is characteristic of substances which have a relatively large overlap of the absorption and fluorescence spectra. The lifetime changes are therefore explained as a reabsorption phenomenon in which the shorter wave-

TABLE 6.6 FLUORESCENCE LIFETIMES AS CALCULATED FROM POLARIZATION AND MOLECULAR VOLUME DATA COMPARED WITH DIRECTLY MEASURED LIFETIMES

Compound	Polarization[a]			Phase shift determinations	
	Solvent	v^b	τ(nsec)	Solvent	τ(nsec)
Fluorescein	H_2O–glycerol	500	4.3	0.01 N NaOH	4.2[c]
Chlorophyll	Cyclohexanol	2500	30	95% EtOH	5.5[d]
Quinine	H_2O–H_2SO_4–glycerol	506	40	N H_2SO_4	20.1[e]
NADH	Water	900	0.35[f]	0.1 M phosphate buffer pH 7.5	0.38[c]

[a]Polarization data from Perrin (1929).
[b]Molecular volumes as determined from viscosity data of Marinesco (1929).
[c]Spencer and Weber (1969).
[d]Muller et al. (1965).
[e]Birks and Dyson (1963).
[f]Weber (1958).

length fluorescence (in the overlap region) is partially absorbed before passing out of the solution. Reemission of this energy then takes place at a time later than the primary emission, but with the same quantum yield and lifetime. Since the secondary emitted quanta must have passed through two excited states and hence two lifetimes, the apparent lifetime for this emission (as measured from excitation of the first molecule to emission from the second) is longer than the primary emission lifetime. Hence, the total lifetime observed is a weighted average of the primary and the secondary lifetimes.

The concentration dependence of fluorescein has been extensively studied (Muller et al., 1965). Table 6.7 lists some lifetimes for fluorescein and for chlorophyll-a as a function of concentration.

If a solution does have a concentration dependence of its fluorescence lifetime, it is possible to check by two methods if it is due to reabsorption. First, as described in Sec. 6.4, when reabsorption is present the short wavelength edge of the emission spectra in a concentrated solution is decreased when compared to dilute solutions. The emission maximum may not be shifted (unless the overlap of absorption and emission extends beyond the wavelength of maximum emission). The quantum yield will be slightly reduced, since some of the primary emitted light is reabsorbed and then remitted with the same quantum yields as for primary emission ($q < 1$). Secondly,

TABLE 6.7 DEPENDENCE OF THE LIFETIME ON CONCENTRATION OF
FLUORESCENT SPECIES

Compound	Concentration	τ (nsec)	Reference
Chlorophyll-a, 95% EtOH	$\sim 3 \times 10^{-7} M$	5.5	a
	$\sim 4 \times 10^{-5} M$	6.5	a
Fluorescein, 0.01N KOH	$10^{-6} M$	3.8	a
	$10^{-4} M$	5.1	a
Fluorescein, 0.01N NaOH	1 μg/ml	4.2	b
	10 μg/ml	4.7	b

[a]Muller et al. (1965).
[b]Spencer and Weber (1969).

since the probability of reabsorption is dependent on both concentration and path length of the solution through which the fluorescent light passes, measurement of lifetimes of concentrated solutions in thin layers should result in values equal to that in dilute solution.

2. *Quenching*

The result of quenching is a decrease in lifetime as well as quantum yield (collisional quenching). This is an obvious source of error in fluorescence measurements, but its importance should be emphasized. Errors of this sort are always possible, because it is simple to forget that the buffer of an aqueous solution may itself quench the molecule under study. A good example of this is the quenching of tyrosine fluorescence by phosphate, a commonly used neutral pH buffer. The quenching of tyrosine in 0.1 M phosphate buffer is significant (Cowgill, 1963) to reduce the lifetime from about 3.4 nsec to 2.6 nsec.

3. *Instrumental Error*

In Tables 6.4, 6.5, and 6.6 we have shown data for the coenzyme nictotinamide adenine denucleotide, NADH. There is a large discrepancy between the data obtained by the pulse method (4.5 nsec) and those by calculation of oscillator strengths, polarization of fluorescence (0.35 nsec), and phase modulation techniques (0.38 nsec). The discrepancy is due to the technical limitations of the pulse techniques. As stated in the first part of the section, limit of resolution is 1 nsec. Thus to measure a compound with a theoretical lifetime of one-third this value is philosophically meaningless if the instrument will not resolve differences in this range. Most probably this value is the result of fluorescent artifacts with lifetimes greater than the NADH. Because of problems of detection, most pulse measurements are made in such a way that one cannot be certain that most of the emission is from the desired compound. To establish that the fluorescence is only from the desired fluorescent species and not from a degradation products requires certain other data such as the fluorescence emission spectrum of the compound.

6.9 Measurement of the Polarization of Fluorescence

A METHODS OF MEASUREMENT

Fluorescence polarization has been extensively discussed in Chapter 4. Recalling the definition of polarization [Eq. (4.1)]:

$$p = (I_{\parallel} - I_{\perp})/(I_{\parallel} + I_{\perp}) \qquad (6.50)$$

it is only necessary to measure the fluorescence intensity emitted parallel to the Z axis ($I_{\parallel}$) and that perpendicular to the Z axis ($I_{\perp}$). Figure 6.22 shows a convenient experimental setup to accomplish this. The plane in which the figure is drawn is the XY plane with the Z axis perpendicular to the plane of the paper.

It shall be assumed that polarized excitation light will be used since, as shown in Chapter 4, the limits on p are $\frac{1}{3}$ to $\frac{1}{2}$, which gives a greater range and sensitivity to measurements than for the case in which natural excitation light is used. The arrangement shown has two analytical arms, A and B, so that $I_{\perp}$ and $I_{\parallel}$ may be measured simultaneously, eliminating any errors to lamp intensity fluctuations between a measurement of $I_{\parallel}$ and $I_{\perp}$. If the polarizer p_A is oriented such that it passes only light which has its electric vector in the XY plane, A will be measuring $I_{\perp}$; if the polarizer P_B orientation passes only light with its electric vector parallel to the Z axis, B will measure $I_{\parallel}$. The excitation polarizer has two positions: (1) passes light in the YZ plane with the electric vector parallel to the Z axis (this is the measuring mode) so as to be perpendicular to P_A and parallel to P_B, and (2) a position rotated by 90° about the Y axis such that the light passed is polarized in the XY plane with the electric vector parallel to the X axis, propagation being along the Y axis. This latter arrangement is referred to as the balance mode because now both P_A and P_B pass light of polarization perpendicular to the exciting light, both are seeing $I_{\perp}$, such that the photomultiplier may be balanced against each other until they give the same output, thus eliminating artifacts arising from mismatched photomultipliers. After balancing, the excitation polarizer

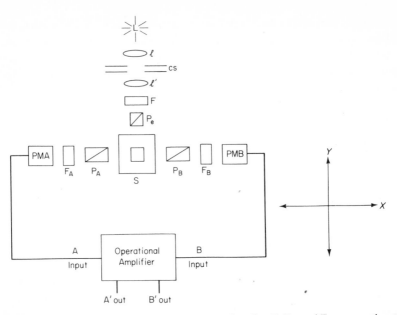

Fig. 6.22. L is lamp; l, l′ are lenses; cs are collimating slits; F, F_A, and F_B are wavelength filters (monochromators could be substituted if desired); P_e is the excitation polarizer which may be revolved 90° about the axis from L to S; S is the sample compartment, usually jacketed for thermal control; P_A and P_B are the fluorescence polarizers for the A and B analytical arms of the instrument; PMA and PMB are the A and B photomultipliers; A, B and A′, B′ are, respectively, the inputs and outputs of the operational amplifier, for details of this see the text; power supplies for the lamp, photomultipliers, and electronic detection systems have not been included.

is returned to its measuring position (electric vector parallel to Z axis) and A and B again see $I_\perp$ and $I_\parallel$, respectively.

Several devices using this arrangement have been designed. An error analysis of one optical arrangement has been presented in great detail (Weber, 1956). An instrument which records the entire polarization excitation spectrum of molecules has been reported (Weber and Bablouzian, 1956).

An instrument which gives an accurate 3-digit readout of p or p^{-1} on a digital voltmeter has been designed by Pasby (unpublished) while Rosén (1971) has described an inexpensive instrument which by use of an integrated-circuit analog divider gives a readout of $I_\perp/I_\parallel$ from which

$$p = \frac{1 - I_\perp/I_\parallel}{1 + I_\perp/I_\parallel}$$

is obtained by simple calculation.

In all cases where filters have been used to isolate a particular wavelength region, monochromators may replace them. This increases spectral resolution at the expense of light intensity. This generally results in higher noise levels due to the necessary increase in photomultiplier voltage and/or amplifier gain in the detection section.

The primary detectors are the photomultipliers which convert the light impinging upon the photocathode into a photocurrent output which is commonly in the range of 10^{-6}–10^{-9} amp. These photocurrents can be directly measured, but it is usually more convenient to convert them to an amplified voltage signal and then record the intensity as potentials with a chart recorder or voltmeter. The operational amplifier labeled in Fig. 6.22 may simply convert the photocurrent to a voltage with subsequent amplification (usually variable) such that A' and B' are directly proportional to $I_\perp$ and $I_\parallel$, and p may thus be calculated by measuring A' and B' and using Eq. (6.50). A more advantageous method is to deliver to A' and B' the sum and difference of $I_\parallel$ and $I_\perp$ by first converting the A and B photocurrents to voltage and then using summing operational amplifiers to take the sum and the difference of the two, thereby delivering to one output (e.g., A'), the sum, $I_\parallel + I_\perp$, and to the other (e.g., B') the difference, $I_\parallel - I_\perp$. p is therefore the ratio B'/A'. An added convenience is to feed A' and B' into either a ratio recorder or a ratio voltmeter, such that p may be recorded directly without calculation.

The above discussion has been for a two-arm analytical instrument ($I_\perp$ and $I_\parallel$ measured simultaneously). It is of course possible to have a one-armed system using only the $I_\parallel$ side of the above described instrument and rotating the excitation polarizer such that the fluorescence photomultiplier sees $I_\perp$ and $I_\parallel$ alternately. The two disadvantages to this are that fluctuations in the exciting light are not compensated, and it is not possible to make direct measurements of p since the sum and difference of $I_\parallel$ and $I_\perp$ and the ratio thereof cannot be displayed as described above.

B. Some Errors in Measurement

The following is a discussion of some of the more common and often significant errors which may give rise to inaccurate polarization data.

1. *Scattering of Exciting Light*

The exciting light may possibly be scattered into the analytical arms of the instrument from cuvette surfaces, cuvette chamber surfaces, and large particles present in the solution under observation.

Light will also be scattered by large molecules in solution (turbidity if visible). In these cases, the polarization of the exciting light is preserved such that it is an added component to $I_\parallel$ but makes no addition to $I_\perp$; we will define this added component as $\delta I_\parallel$ and remember that it is of the wavelength of excitation, λ_e, whereas $I_\parallel$ and $I_\perp$ are of the wavelengths of fluorescence, λ_F. If the fluorescence filters or monochromators used to isolate the fluorescence band of the fluor effectively stop all light of wavelength λ_e,

$$p = (I_\parallel - I_\perp)/(I_\parallel + I_\perp) \qquad (6.51a)$$

but if some λ_e passed, then

$$p' = (I_\parallel - c\,\delta I_\parallel - I_\perp)/(I_\parallel + c\,\delta I_\parallel + I_\perp) \qquad (6.51b)$$

and $p' > p$. c is a constant to account for the difference in transmission of the filter for λ_e to λ_F and also the different response of the photomultiplier for λ_e and λ_F.

This effect becomes very important when the polarization is very low, $I_\parallel - I_\perp$ is small, such that $c\,\delta I_\parallel$ becomes proportionately greater in total contribution to the numerator. There is no easy way to check for this artifact, especially in precise terms as to the exact size of $c\,\delta I_\parallel$. If there is suspicion that light scattering is contributing to p, change emission filters such that λ_F is further separated from λ_e and note if p decreases. If a monochromator is used, using narrower slits as well as going to longer λ_F's may be tried. To establish approximate limits as to the size of $c\,\delta I_\parallel$, a maximum size may be obtained by measuring $I_\parallel$ of a nonfluorescing, scattering solution, such as glycogen, which has a turbidity at λ_e equal to the OD of the fluor solution at λ_e. To determine scattering which arises from solvent, cuvette surfaces, and compartment surfaces, $I_\parallel$ may be measured on a solvent blank. None of these can give an absolute measure of $c\,\delta I_\parallel$, but they can establish if scattering may be contributing to the measured polarization.

2. High Absorbance of Fluor

When the absorbancy of the solution at λ_e is too large, all absorbance and hence fluorescence may be occurring at or very near the front of the cuvette. Since most sample compartments are designed not to observe this portion of the cuvette so as to decrease any observed scatter from this cuvette surface, the fluorescence may also not be directly observed and the only fluorescence reaching the photomultiplier is from light which has been reflected off cuvette surfaces an unknown number of times before finding its way down the optical path to the emission photomultiplier. This radiation has an ill-defined relationship to the initial emitted radiation and the p value obtained may or may not be valid.

If there is reason not to change the absorbancy by dilution (the obvious remedy) then an alternative is to have a shorter path length of solution. One way is use of cuvettes of smaller path length (e.g., changing from a 1 cm × 1 cm cuvette to a 0.1 cm × 0.1 cm cuvette) which is then placed in the cuvette compartment such that the fluorescence is directly observed. Another method is to use prisms between which the solution is placed producing a very thin layer of solution and hence a very short path length, allowing observation from the center of the sample compartment.

3. Trivial Reabsorption

When there is large overlap between fluor absorption spectrum and emission spectrum, some of the emitted light may be reabsorbed and reemitted before passing out of the cuvette; this light coming from a secondary (or tertiary, etc.) emission process has undergone two stages of depolarization, and, therefore, the measured polarization is less than if no reabsorption occurs. When this happens, the only recourse is to use more dilute solutions or shorter pathlengths, or both.

4. Intrinsic Flourescence of Color Filters

This is related to absorbancy, for in many cases fluorescence is observed using color filters which absorb light at wavelengths $< \lambda_F$ but pass light of wavelength $> \lambda_F$, thus, λ_e is absorbed and λ_F transmitted. However, some color filters have an intrinsic luminescence (fluorescence or phosphorescence) excited by the excitation light λ_e. Therefore, even though scattered λ_e may never reach the photomultiplier, $\lambda_{luminescence}$ may reach the photomultiplier and will thus be interpreted as a component of $I_{\parallel}$. In most practical cases, λ_e is in the ultraviolet and λ_F is visible. If this is the case, a filter consisting of a 2-mm thickness of $2\,M$ $NaNO_2$ may be placed on the sample side of the color filter; the $NaNO_2$ is opaque to UV below about 380 nm but is transparent to visible light.

REFERENCES

Bauer, R., and Rozwadowski, M., *Bull. Acad. Polon. Sci.*, CIII, **7**, 365 (1959).
Birks, J. B., and Dyson, D. J., *J. Sci. Instr.*, **38**, 382 (1961).
Birks, J. B., and Dyson, D. J., *Proc. Roy. Soc. (London)*, **A275**, 135 (1963).
Birks, J. B., and Monroe, T. H., *Progr. Reaction Kinetics*, **4**, 239 (1967).

Bowen, E. J., and Sawtell, J. W., *Trans. Faraday Soc.*, **33**, 4425 (1937).

Bowen, E. J., and Williams, A. H., *Trans. Faraday Soc.*, **35**, 765 (1939).

Chen, R. F., Vurek, G. G., and Alexander, N., *Science*, **156**, 949 (1967).

Cowgill, R. W., *Arch. Biochem. Biophys.*, **100**, 36 (1963).

D'Allesio, J. T., Ludwig, P. K., and Burton, M., *Rev. Sci. Instr.*, **35**, 1015 (1964).

Dushinsky, F., *Z. Physik.*, **81**, 7 (1933).

Einstein, A., *Z. Physik.*, **18**, 121 (1917).

Ellis, D. W. and Rogers, L. B., *Spectrochim Acta*, **18**, 263 (1962); **20**, 1709 (1964).

Förster, T., *Fluoreszenz, Organischer Verbindungen*, Vandenhoeck and Ruprecht, Gottingen, 1951.

Forster, L. S., and Livingston, R., *J. Chem. Phys.*, **20**, 1315 (1952).

Gaviola, E., *Z. Physik.*, **42**, 852 (1927).

Gilmore, E. H., Gibson, G. E., and McClure, D. S., *J. Chem. Phys.*, **20**, 829 (1952).

Gilmore, E. H., Gibson, G. E., and McClure, D. S., *J. Chem. Phys.*, **23**, 399 (1955).

Hatchard, C. G., and Parker, C. A., *Proc. Roy. Soc. (London)*, **A235**, 518 (1956).

Hercules, D. M., and Frankel, H., *Science*, **131**, 1611 (1960).

Hundley, L., Coburn, T., Garwin, E., and Stryer, L., *Rev. Sci. Instr.*, **38**, 488 (1967).

Knopp, J. A., Ph.D. Thesis, University of Illinois, Division of Biochemistry, 1967.

Kostkowski, H. J., and Lee, R. D., *Natl. Bur. Stds. Monograph* 41, March, 1962.

Lewis, G. N., and Kasha, M., *J. Am. Chem. Soc.*, **67**, 994 (1945).

Maerks, O., *Z. Physik.*, **109**, 685 (1938).

Marinesco, N., *J. Chim. Phys.*, **24**, 593 (1929).

McLachlan, N. W., *Complex Variable Theory and Transform Calculus*, 2nd ed., Cambridge Univ. Press, London and New York, 1963, p. 184.

Melhuish, N. Z., *J. Sci. Tech.*, **B37**, 142 (1955).

Melhuish, N. Z., *J. Phys. Chem.*, **64**, 762 (1960).

Melhuish, N. Z., *J. Phys. Chem.*, **65**, 229 (1961).

Melhuish, N. Z., *J. Opt. Soc. Am.*, **52**, 1256 (1962).

Melhuish, N. Z., *J. Opt. Soc. Am.*, **54**, 183 (1964).

Miller, G., Johnson, J. A., and Miller, B. S., *Anal. Chem.*, **28**, 884 (1956).

Muller, A., Lumry, R., and Kokubun, H., *Rev. Sci. Instr.*, **36**, 1214 (1965).

Parker, C. A., *Proc. Roy. Soc. (London)*, **A220**, 104 (1953).

Parker, C. A. and Barnes, W. J., *Analyst*, **82**, 606 (1957).

Parker, C. A., *Nature*, **182**, 1002 (1958).

Parker, C. A., *Analyst*, **84**, 446 (1959).

Parker, C. A., and Rees, W. T., *Analyst*, **85**, 587 (1960).

Parker, C. A., and Rees, W. T., *Analyst*, **87**, 83 (1962).

Perrin, F., *J. Phys. Radium*, **7**, 390 (1926).

Perrin, F., *Ann. Phys.*, **12**, 169 (1929).

Rosén, C.-G. (1971). *Acta Chem. Scand.* On press.

Shepp, A., *J. Chem. Phys.*, **25**, 579 (1956).

Spencer, R. D., and Weber, G., *Ann. N.Y. Acad. Sci.*, **158**, 361 (1969).

Spencer, R. D., and Weber, G., *J. Chem. Phys.*, **52**, 1654 (1970).

Stair, R., Schneider, W. E., and Jackson, J. K., *Appl. Opt.*, **2**, 1151 (1963).

Stevens, B., and Hutton, E., *Nature*, **186**, 1045 (1960).

Strickler, S. J., and Berg, R. A., *J. Chem. Phys.*, **37**, 814 (1962).

Strickler, H. S., Grauer, R. C., and Caughey, M. R., *Anal. Chem.*, **28**, 1240 (1956).

Stryer, L., *Science*, **162**, 526 (1968).

Umberger, J. Q., and LaMer, V. K., *J. Am. chem. Soc.*, **67**, 1099 (1945).

Vavilov, S. I., *J. Physik.*, **22**, 266 (1924).

Ware, W. R., and Baldwin, B. A., *J. Chem. Phys.*, **40**, 1703 (1964).
Weber, G., *J. Opt. Soc. Am.*, **46**, 962 (1956).
Weber, G., *J. Chem. Phys.*, **55**, 878 (1958).
Weber, G., *Nature*, **190**, 27 (1961).
Weber, G., and Bablouzian, B., *J. Biol. Chem.*, **241**, 2558 (1956).
Weber, G., and Teale, F. W. T., *Trans. Faraday Soc.*, **53**, 646 (1957).
Weber, G., and Teale, F. W. T., *Trans. Faraday Soc.*, **54**, 640 (1958).
Weber, G., and Young, L. B., *J. Biol. Chem.*, **239**, 1424 (1964).
White, C. E., Ho, M., and Weimer, E. Q., *Anal. Chem.*, **32**, 438 (1960).

BIBLIOGRAPHY

Bowen, E. J., *The Chemical Aspects of Light*, 2nd ed., Clarendon Press, Oxford, 1946.
Hercules, D. M., *Fluorescence and Phosphorescence Analysis*, Wiley (Interscience), New York, 1966.
Jaffe, H. H., and Orchin, M., *Theory and Applications of Ultraviolet Spectroscopy*, Wiley, New York, 1962.
Pringsheim, P., *Fluorescence and Phosphorescence*, Wiley (Interscience), New York, 1949.
Udenfriend, S., *Fluorescence Analysis in Biology and Medicine*, Academic Press, New York, 1962.

Chapter 7

Use of Fluorescence in Binding Studies

Introduction

The use of fluorescence spectroscopy for the study of the binding properties of a number of proteins and nucleic acid polymers has grown at a fast rate. Much information concerning the nature of protein binding mechanisms and the intrinsic structure of proteins has come out of this effort. This chapter is intended to be a guide to the use of fluorescence as a mode of measuring ligand–polymer interactions.

7.1 Study of Protein Binding

Many recent studies have been concerned with the interaction of proteins with small molecules.* The binding of small molecules and related protein–protein interactions and conformational changes are important to the biological activity of proteins. Examples of such systems are enzymes, which bind substrates, effectors, and prosthetic groups, gas binding respiratory proteins, metabolite binding transport proteins, electron binding oxidation-reduction enzymes, and antibody–antigen systems. The ability of many proteins to bind nonbiological "artificial probes," which are designed to mimic natural ligands or to bind to hydrophobic protein regions, has also been utilized in studying protein binding.

Recent work has led to an understanding of the specificity with which a protein binds a ligand in terms of a site on the protein which

*We shall refer to such small molecules as ligands, using the general terminology for molecular complex formation.

may interact with the ligand through ionic forces, hydrogen bonds, or hydrophobic forces. The same sort of specific physical interactions involved in protein binding may be found in the organization of other cellular components, but proteins appear unique in the variety, specificity, and control of interactions. The complex nature of protein–small molecule interactions, and the possibility of studying them as isolated physical systems, makes them interesting and important. Studies of the mechanisms of protein interactions and conformational changes provide insight into the dynamics of protein action, and may contribute to understanding many of the mechanisms underlying biological specificity.

Binding may be studied by both kinetic and equilibrium methods, and the data obtained by these methods are often complementary. Kinetic methods have been widely used to study binding steps in enzyme reactions (Vestling, 1963; Day et al., 1963), and the binding of allosteric effectors to enzymes. Recently, systematic consideration has been given to equilibrium methods and the information obtainable from them (Weber, 1965; Weber and Anderson, 1965). This discussion will be limited largely to such equilibrium studies.

Characterization of binding equilibria involves the determination of concentrations of components of the equilibrium system; i.e., P, L, and PL in the equilibrium $P + L \rightleftharpoons PL$. This is essentially the same thing that is done in a simple acid-base titration, in which the concentration of free ligand (H^+) and the amount of acid or base added are determined. A similar titration curve characterizes the binding equilibrium of a protein–small molecule system and an acid-base system. See Fig. 7.1. We will see that, in the complex binding often found in proteins, such a complete titration curve is capable of yielding much more information than a simple equilibrium constant.

The titration curve may be plotted in several ways, but one of the most useful is analogous to the common acid-base plot: negative logarithm of the free ligand concentration vs. saturation of the binding species.

Fluorescence spectroscopy is one of the most versatile and sensitive methods for physical studies of protein systems. It has been used to follow protein denaturation, conformational transitions, and changes in subunit association, to characterize binding processes, and to measure the polarity of binding sites and the orientation and distribution of ligands. Such studies are discussed in review articles by Velick (1961), Steiner and Edelhoch (1962), Chen (1967a), Parker (1967), Edelman and McClure (1968), and Stryer (1968). Less attention has been given to the study of actual binding processes, for which fluorescence is also

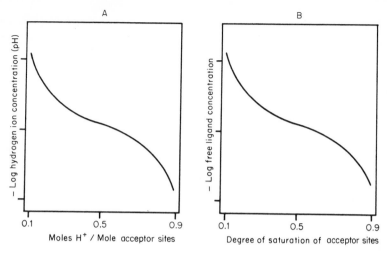

Fig. 7.1. Comparison of a pH titration curve (A) with a simple ligand binding curve (B).

a powerful method. Fluorescence measurements can be made under a wide variety of conditions, and large amounts of information can be obtained. Fluorescence methods, while complementary to other physical and chemical methods, often provide information not obtainable in other ways.

7.2 Measurement of Binding System Parameters

In order to characterize a binding system we would like to determine such things as the number of binding sites, their affinities, nature, location on the protein, and the biological significance of their binding (Edsall and Wyman, 1958). In addition, much can often be learned about the mechanism of binding. We will deal here with the stoichiometry and equilibria of binding, and their implications for elucidating binding mechanisms. These parameters may be described in detail mathematically and measured with great precision by fluorescence methods.

Several comprehensive discussions of binding equilibria have been published, among which are Klotz (1953), Bjerrum (1957), Edsall and Wyman (1958), Tanford (1961), Jaffe and Orchin (1962), Weber (1965), and Deranleau (1969). Since an understanding of binding theory is a necessary guide to experimental design and interpretation, some of the more relevant theory will be reviewed here, with a specific view toward protein binding systems.

We will first derive some general equations that describe binding, and then use them to discuss the determination of stoichiometry and binding equilibria.

A. GENERAL BINDING EQUATIONS

Taking a simple equilibrium between a protein (P) and one molecule of ligand (L), we may write

$$P + L \; \rightleftharpoons \; PL \qquad\qquad (7.1)$$

with the association constant* of the system, K_a, related to the concentrations of the components:

$$K_a = \frac{[PL]}{[P][L]} \qquad\qquad (7.2)$$

For simplicity we shall deal with a protein having only one binding site, so that the concentration of possible binding sites is equal to $[P_0]$, the total protein concentration. Then we have

$$\phi = \frac{\text{Occupied sites}}{\text{Possible sites}} = \frac{[PL]}{[P]+[PL]} = \frac{[PL]}{[P_0]} \qquad\qquad (7.3)$$

where ϕ is the fractional saturation of the protein, and $[P]$ is the concentration of unliganded protein.

By rearranging Eq. (7.2) and substituting Eq. (7.3) into it, we can express K_a in terms of ϕ:

$$K_a = \frac{[PL]/[P_0]}{[P][L]/[P_0]} = \frac{\phi}{1-\phi} \frac{1}{[L]} \qquad\qquad (7.4)$$

A derivation of (7.4) perhaps more familiar to some readers is that given by Edsall and Wyman (1958): rearranging Eq. (7.2) and substituting it into Eq. (7.3) we get an equation identical to (7.4), namely,

$$\phi = \frac{[PL]}{[P]+[PL]} = \frac{K_a[L]}{1+K_a[L]} \qquad\qquad (7.4a)$$

---

*The association and dissociation constants are not two constants, but rather one equilibrium constant which may be written for a reaction proceeding in either direction, so that the two are simply the inverse of one another.

When the equilibrium constant is defined by concentrations instead of activities it is not a true constant, but is solvent-dependent, and is referred to as an *apparent constant*. However, in most experiments, at approximately constant ionic strength, the apparent constants usually differ from the true constants by less than experimental error.

More obvious meaning may be given to Eq. (7.4) by picturing $K_a[L]$ as the ratio of occupied sites to empty sites, $\phi(1-\phi)$. We may rearrange (7.4) as follows:

$$\frac{1}{K_a}\frac{\phi}{1-\phi} = [L] = [L_0] - [PL]$$

where $[L_0]$ equals the total ligand concentration, $[L] + [PL]$. Dividing by $[P_0]$,

$$\frac{1}{K_a[P_0]}\frac{\phi}{1-\phi} = \frac{[L_0]}{[P_0]} - \frac{[PL]}{[P_0]} = \frac{[L_0]}{[P_0]} - \phi$$

and rearranging, we get

$$\frac{[L_0]}{[P_0]} = \frac{1}{K_a[P_0]}\frac{\phi}{1-\phi} + \phi \tag{7.5}$$

From this equation we can obtain the relationship of the amount of ligand added, $[L_0]$, to the fractional saturation, ϕ, at constant protein concentration. This is precisely what is done in a normal titration, in which ligand is added to a protein solution, and ϕ is measured.

The possibility of complexes with more than one protein molecule will not be considered here, but effects of protein-protein interactions on binding may be quite important, and will be discussed in Sec. 7.6C.

B. STOICHIOMETRY

The stoichiometry of binding is the maximum number of ligand molecules (N) a protein will bind. The number of binding sites in a system must be determined before the binding affinity (equilibrium constant) can be calculated. It is quite possible that a system will have several distinct classes of binding sites, where a class consists of a set of identical sites (i.e., of identical affinities). We may be concerned with a set of relatively strong and specific sites, to the exclusion of a set of weaker sites. If the sites are of sufficiently different affinity, the weaker ones will be essentially unoccupied under conditions in which the stronger sites are binding, and may be ignored. As two classes of sites approach one another in affinity it becomes more difficult to study them separately.

Stoichiometry and equilibrium are both determined by a simple titration, but they are best measured under different conditions. Stoichiometry must, in general, be determined under conditions in which all added ligand is fully bound to the protein, with no detectable amount free—i.e., stoichiometrically bound.* Binding *equilibrium* information must be obtained under conditions in which an equilibrium obtains, so that there are detectable amounts of all three components of the system: free protein, free ligand, and protein–ligand complex.

In order to clarify the experimental methods involved, we will first derive the conditions under which stoichiometric binding can be measured, by showing the dependence of binding on protein and ligand concentration. We will show that when $[P_0]$ is much greater than K_d, added ligand binds virtually stoichiometrically to the protein, allowing us to infer the amount bound at saturation by its equality to the amount added at saturation, in a titration-type experiment.

Equation (7.5) shows ϕ is dependent on $[P_0]$, and before proceeding further with a discussion of the conditions of stoichiometry, the dependence should be understood: that Eqs. (7.4) and (7.5) also hold for N binding sites, in the equilibrium $P + NL \rightleftharpoons PL_N$, as long as the sites are equivalent and independent. Therefore we may use these equations to discuss cases with a stoichiometry greater than one.

Let us introduce the fraction of ligand bound, f, which depends on $[P_0]$, as we may see by writing

$$f = \frac{[L_{bound}]}{[L_{total}]} = \frac{[L_0] - [L]}{[L_0]} = \frac{[PL]}{[L_0]} = \frac{[PL]/[P_0]}{[L_0]/[P_0]}$$

and, from (7.5),

$$f = \phi \Big/ \left[\frac{1}{K_a[P_0]}\left(\frac{\phi}{1-\phi}\right) + \phi \right] \tag{7.6}$$

If $[P_0] \gg 1/K_a$, the first term in the denominator becomes small and f approaches 1, or all the ligand in solution is stoichiometrically bound. In this case, the equation reduces to two linear equations, $\phi = [L_0]/[P_0]$ at low $[L_0]$, and $\phi = 1$ at $[L_0]$ approaching $[P_0]$ or saturation. The equation $\phi = [L_0]/[P_0]$ describes stoichiometric binding of the ligand, i.e., increasing $[L_0]$ results in a linear increase in ϕ: at high enough $[P_0]$ all of the ligand binds stoichiometrically, as long as there are sites available on the protein. Even under the so-called stoichiometric condition of $[P_0] \gg 1/K_a$, Eq. (7.1) is valid and there is an equilibrium, but it is undetectably small; the concentration of free ligand is experimentally indistinguishable from zero. The concentration of free ligand which is just below the limits of detectability or

*Note the distinction between "the stoichiometry of binding," the number of sites on the protein, and "stoichiometric binding," in which either the protein or ligand (or both) is fully bound.

acceptability in a given situation determines the minimum by which $[P_0]$ must exceed $1/K_a$ in order to obtain stoichiometric conditions.

It is often not possible in protein binding systems to achieve a protein concentration which is much greater than $1/K_a$, since many systems have K_a's of the order of $10^5\ M^{-1}$ or lower. (In practice, a factor of 10 greater than $1/K_a$ is usually sufficient for $[P_0] \gg 1/K_a$.) When $1/[P_0]$ comes close enough to K_a for the first term on the right of Eq. (7.6) to become significant, a detectable equilibrium exists, and the equation is no longer reducible to two linear segments. The straight-line segments begin to curve as shown in Figs. 7.2 and 7.3, where we have plotted the equation as $[L_0]/[P_0]$ vs. ϕ. The region of curvature, where the term

$$\frac{1}{K_a[P_0]}\left(\frac{\phi}{1-\phi}\right)$$

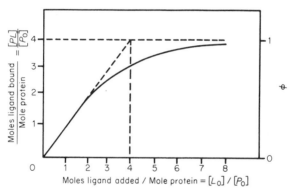

Fig. 7.2. Stoichiometric binding curve for a case of $N = 4$.

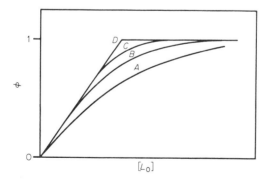

Fig. 7.3. Binding surface, showing the dependence of binding on $[P_0]$ and $[L_0]$ in the absence of any anomalous protein concentration efforts. Curves A through D are for increasing protein concentration.

is significant, is the region of detectable equilibrium in which all three components, P, PL, and L are present in appreciable amounts. By the same reasoning, it is seen that the curvature becomes greater, or the linear region becomes smaller, as K_a decreases with $[P_0]$ held constant, a situation which might arise for a series of analogs which bind to a protein with different affinities.

We may also picture the dependence of binding on $[P_0]$ in a slightly different way. We redraw Fig. 7.2 for several increasing values of $[P_0]$, as shown in Fig. 7.3. The result is essentially a contour map, in which we may picture the $[P_0]$ axis at 90° to the other two. The three-dimensional surface formed by joining the lines is called the binding surface (Weber, 1965). We have seen that the surface varies in a predictable way with $[P_0]$ when Eq. (7.5) is obeyed. However, we will see that there are many cases in which changes in $[P_0]$ actually change the form of the binding curve, by subunit dissociation or other concentration-dependent perturbations. In such cases, it is necessary to characterize the binding surface in some detail by studying the dependence of binding equilibria on $[P_0]$ or, if possible, to confine measurements to a range of $[P_0]$ in which the concentration-dependent effects are negligible.

C. EQUILIBRIA

The preceding discussion on stoichiometry has already introduced equilibrium binding conditions: $[P_0]$ approximately equal to $1/K_a$. Under these conditions we wish to characterize binding equilibrium. In the simplest case this amounts to determining "the equilibrium constant," but in more complex cases it is better to say we wish to determine ϕ as a function of $[L]$. We will see that it is not always possible to determine a simple K_a. The dependence of the degree of saturation on $[L]$ is straightforward. We can rewrite (7.4) as

$$\frac{[L]}{K_d} = \frac{\phi}{1-\phi}$$

where we have chosen to return to the dissociation constant. (It is more straightforward to use the association constant to write equations, but in thinking about binding equilibrium the dissociation constant has more intuitive meaning since its units are those of ligand concentration, instead of the reciprocal.) We see that at $[L] = 10K_d$, $\phi = 0.91$, and at $[L] = 0.1K_d$, $\phi = 0.091$. So ϕ varies from about 0.1 to 0.9 as $[L]$ varies from 10-fold less than K_d to 10-fold greater than K_d.

Equation (7.4) may be rearranged and put in log form, which has certain advantages in data plotting (see the Appendix):

$$\log K_a + \log [\text{L}] = \log \left(\frac{\phi}{1-\phi}\right) \tag{7.7}$$

We may again point out the analogy between the binding equations derived here and the equations describing acid-base equilibria, which are the basis for acid-base titrations. We may rearrange Eq. (7.7) (remembering that $\log n = -\log 1/n$) and obtain the equivalent of the well-known Henderson-Hasselbalch equation:

$$\log [\text{L}] = -\log K_a + \log \left(\frac{\phi}{1-\phi}\right)$$

$$-\log [\text{L}] = \log K_a + \log \left(\frac{1-\phi}{\phi}\right)$$

$$-\log [\text{H}^+] = -\log K_a + \log \frac{\text{unprotonated form}}{\text{protonated form}}$$

$$\text{pH} = \text{p}K_a + \log \frac{\text{unprotonated form}}{\text{protonated form}}$$

To evaluate binding equilibria we must make measurements on a system with a measurable concentration of all three components—free protein, free ligand, and protein-ligand complex. In order to obtain a system of detectable equilibrium, unlike the conditions of stoichiometric binding, we must have $[\text{P}_0]$ less than or about equal to K_d, and certainly not greater than a factor of 10 different. In this case, the plot of $[\text{L}_0]/[\text{P}_0]$ vs. ϕ has almost no initial stoichiometric segment and reaches saturation at a higher $[\text{L}_0]$ because the range of saturation over which there is appreciable free ligand is much greater than for a higher $[\text{P}_0]$. Since the approach to saturation is more shallow it is difficult to determine stoichiometry under equilibrium conditions (just as it is difficult to determine equilibrium behavior under stoichiometric conditions). However, the two conditions are merely limits of behavior of the curve, with a continuum of possible curves in between, and it is therefore possible that K and N can be determined from the same curve, with some sacrifice of accuracy. The different conditions merely emphasize the different information needed for the two cases and make it more clearly obtainable by graphic means.

7.3 Binding Systems Amenable to the Use of Fluorescence Methods

As discussed in Sec. 7.1, fluorescence methods for measurement of binding equilibria depend on the change of a measurable parameter of the protein-ligand system with binding. The parameter most commonly utilized is the fluorescence intensity of the protein or ligand; further information can be gained from changes in the fluorescence polarization, which will be discussed in Sec. 7.6.

Several conditions must be satisfied for a binding system to be amenable to fluorescence measurements. The change in fluorescence should be rapid relative to the binding process, so that the only time-dependent processes (if there are any on the time scale of a series of fluorescent measurements) are actual binding processes. In addition, the dissociation constant of a ligand cannot be less than approximately $10^{-8} M$. This limit is imposed by the concentration of bound ligand required to give a readable fluorescence signal with normal light sources and detectors, and also by the lowest concentrations of proteins that can generally be handled without denaturation (see Sec. 7.6A).

In cases in which the ligand itself undergoes a change in fluorescence on binding, it is commonly called a *fluorescent probe* or *fluorescent dye*. Compounds which meet the requirements of a measurable fluorescence change on binding, appropriate affinity, and interesting specificity may seem rather rare. In actuality many such compounds are known, among which are natural prosthetic groups such as flavins and pyridoxal phosphate, substrate analogs which may or may not undergo reaction, and artificial fluorescent dyes which bind largely to hydrophobic protein surfaces. (Many fluorescent compounds involve relatively rigid aromatic structures which, while they must be soluble in water to be useful, have an affinity for protein hydrophobic regions).

One may also study binding systems in which the ligand itself does not undergo fluorescence changes on binding, but some group on the protein does, such as a covalently attached fluorescent probe, a prosthetic group, or the inherent protein fluorescence. It may be difficult however to obtain assurance that a covalently attached probe does not perturb the binding behavior of the protein.

In systems in which binding is specific to one particular protein, it may be studied in the presence of other proteins. If they interact with the binding protein the interaction may often be studied by this technique [cf. Rawtich and Weber (1969)].

Table 7.1 lists some representative studies which may serve as examples of the broad range of applications of fluorescent binding studies.

TABLE 7.1 SOME LIGAND–PROTEIN BINDING STUDIES[a]

Protein	Ligand	Reference
A. Fluorescent Dye as the Ligand		
Bovine serum albumin	Naphthalene sulfonates	Laurence (1952)
		Weber and Young (1964)
		Daniel and Weber (1966).
	Rhodaminine, fluoresceine, acridine derivatives	Laurence (1952)
Lactate dehydrogenase	1,8-ANS[b]	Anderson and Weber (1966).
Alcohol dehydrogenase	ANS isomers	Brand and Gohlke (1966)
	Naphthalene sulfonates and rose bengal	Brand et al. (1967)
α-Chymotrypsin and	1,5-DNS[c]	Deranleau and Neurath (1966)
chymotrypsinogen	2,6-TNS[d]	McClure and Edelman (1967)
Antibodies	Derivatives of 2,4-dinitrophenyl and p-azo-phenyl-arsonate	Velick et al. (1960)
	ANS and TNS derivatives	Winkler (1962)
	ϵ-DNP-lysine	Eisen and Siskind (1964)
	1,4–ANS	Parker and Osterland (1966);
		Yoo et al. (1967)
	ϵ-DNS-lysine	Parker (1967)
Bence–Jones protein	1,4-ANS	Gally and Edelman (1965)
Various proteins	Dansyl-amino acids	Chen (1967b)
B. Fluorescent Coenzyme as the Ligand		
Alcohol dehydrogenase	NADH	Boyer and Theorell (1956)
		Theorell et al. (1961)
Lactate dehydrogenase	NADH	Velick (1958)
		Winer et al. (1959)
		Winer and Schwert (1959)
		Vestling (1963)
		Fromm (1963)
		McKay and Kaplan (1964)
		Anderson and Weber (1965)
Glutamic aspartic transaminase	Cycloserine	Churchich (1967)
Dihydrofolate reductase	NADPH, NADP[+], folate and dihydrofolate analogs	Perkins and Bertino (1966)
Malic enzyme	NADP, NADPH	Hsu and Lardy (1967)
C. Covalent Lable or Inherent Protein Fluorescence		
Aphohemoglobin/heme	DNS chloride	Weber and Teale (1959).
Trypsin/trypsin inhibitor	DNS chloride	Millar et al. (1962)
Bovine serum albumin fragment association	DNS chloride	Weber and Young (1964)
Bence–Jones protein	DNS chloride	Gally and Edelman (1965)
α-Chymotrypsin	Anthraniloyl group	Haugland and Stryer (1967)
Bovine serum albumin/steroids	Protein fluorescence	Attalah and Lata (1968)
Alumbin/thyroxine	Protein fluorescence	Steiner et al. (1966)
Transferrin/Fe[3+] and Cu[2+]	Protein fluorescence	Lehrer (1969)
Antibodies/haptens	Protein fluorescence	Day et al. (1963)
Cytochrome b_2/heme	Protein fluorescence	Labeyrie et al. (1967)

[a] For further examples, see the review articles cited in Sec. 7.2.
[b] ANS – anilinonaphthalene sulfonate
[c] DNS – dimethylaminonaphthalene sulfonate
[d] TNS – toluidylnaphthalene sulfonate

7.4 Experimental Determination of Binding

A. Methods of Titration

We pointed out in the previous section that stoichiometry and equilibrium are both determined by titration procedures, under different conditions. Something often overlooked in binding studies is the importance of the method of doing the titration. There are two basic methods of doing a titration—an addition method with essentially constant protein concentration, and a dilution method with the protein concentration varying with $[L_0]$ from stoichiometric concentrations through the whole equilibrium region. Knowledge of the advantages and limitations of each method is very important for protein binding studies.

1. Addition Titration

This method, analogous to the common acid-base titration, consists of adding successive increments of ligand to a protein solution. Ideally the volume of ligand added should be small, so that the protein concentration will remain almost constant. It is therefore necessary to do separate addition titrations to determine stoichiometry and equilibria. In certain cases, time-dependent changes may occur in the protein during binding studies, and it will be necessary to use a new protein solution with successively increasing amounts of ligand for each point on the titration curve.

This method has the advantage that with a readily soluble ligand the protein concentration can be kept in a relatively narrow range, but care must be taken that the protein concentration is low enough that the first part of the binding curve is not lost to stoichiometric binding.

For ligands with an association constant of about $10^6\,M^{-1}$ or more it is difficult to avoid losing much of the binding curve to stoichiometric conditions. Many reports in the literature have been based on the addition titration method, with very few points in the equilibrium region and no information on the binding process except an average binding constant.

2. Dilution Titration

A method much superior, particularly for association constants greater than about $10^6\,M^{-1}$ to $10^7\,M^{-1}$, is the dilution method (Anderson and Weber, 1965). A dilution titration begins with an initial solution of protein and ligand, in which the concentration of ligand is slightly greater than the concentration of protein binding sites,* and of sufficient concentration that ϕ is about 0.9 to 0.95 (i.e., a protein and ligand concentration at least 10-fold greater than the dissociation constant of the complex). It is usually not possible to detect binding changes with sufficient accuracy in the first and last 10% of saturation, so there is no advantage in beginning the titration at a ϕ above 0.9 to 0.95. The solution is diluted in a series of steps spaced so as to obtain the most information, i.e., the greatest resolution of the binding curve. In a good titration one should be able to obtain at least 20 to 30 points, spaced at about twice the standard deviation of a single point. The dilution should go, if possible, to $\phi = 0.1$ or 0.05, or about 100-fold from the initial solution (10-fold on each side of the average dissociation constant). If it is not possible to begin at a saturation as great as 0.9, one may reach higher values of saturation for the purpose of measuring stoichiometry, with a separate addition titration starting with the same initial solution.

With a dilution titration, $[P_0]$ begins in the stoichiometric region, so stoichiometry and equilibrium may be determined from the same experiment. It is necessary in this method to show that protein–protein interactions or subunit dissociations do not affect ligand binding over a protein concentration range of at least 10-fold on either side of the average dissociation constant (or the highest and lowest constants in the case of multiple binding).

Binding equilibrium behavior (i.e., K) should be independent of $[P_0]$, although the range of saturation in which stoichiometric binding of the added ligand occurs is dependent on $[P_0]$. However, normal binding equilibrium may be perturbed in protein systems by protein–protein interactions, which are dependent on the protein activity coefficient (or, for our purposes, concentration). It is necessary to establish that, in the range of protein concentrations used in a binding study, binding is independent of the protein concentration. Conversely, in proteins with subunits it is possible to reverse the situation and study the effects of subunit dissociation on binding. In this latter case one must obtain an independent measurement of dissociation, while in the former case the demonstration of any dependence upon protein concentration is sufficient. In Section 7.6C we will discuss the experimental determination of the range of protein concentrations in which binding behavior is not perturbed.

Several recent papers have discussed protein–protein equilibria. For details, see Deranleau (1964), Chun and Fried (1967), Nichol et al. (1967), Adams and Lewis (1968), and Klapper and Klotz (1968).

*The protein concentration is limiting and the definition of ϕ is still $[PL]/[P_0]$ instead of $\phi = [PL]/[L_0]$.

7.5 Calculation of Binding Parameters from Fluorescence Measurements

Binding parameters (concentrations of components of an equilibrium, and therefore equilibrium constants) can easily be calculated from fluorescence *intensity* measurements, for ligand or protein fluorescence changes on binding, whether one is using either inherent protein fluorescence or a covalent fluorescent label. Since the methods depend on the change of intensity on binding, it is necessary to know the intensity characteristics of the fully free state and the fully bound state, which we will call F_f and F_b, respectively. We will now define the following:

F_{sample} = fluorescence intensity of a protein–ligand solution
$\quad F_{std}$ = fluorescence intensity of an appropriate standard (a fluorescent solution of the same absorbance as the sample at the exciting wavelength. See Sec. 7.6A for a discussion of the use of fluorescence standards)
$\quad F_{rel} = F_{sample}/F_{std}$ = fluorescence yield* of a given sample
$\quad F_f$ = fluorescence yield of *free* ligand or protein = $F_{free\,ligand}/F_{std}$
$\quad F_b$ = fluorescence yield of *fully bound* ligand or protein = $F_{bound\,ligand}/F_{std}$

The fluorescence yield of a protein–ligand mixture, in which $[P_0]$ and $[L_0]$ are known, allows the calculation of all three components of the equilibrium. We may calculate the fraction of ligand bound by taking the fluorescence yield of a mixture of L and PL as the sum of the yields of its components.

$$F_{rel} = F_f[L] + F_b[PL]$$
$$= F_f([L_0] - [PL]) + F_b[PL] \qquad (7.8)$$
$$= F_f[L_0] + (F_b - F_f)[PL]$$

Dividing by F_f we get

$$\frac{F_{rel}}{F_f} = 1 + \frac{(F_b - F_f)[PL]}{F_f[L_0]}$$

But $[PL]/[L_0] = f$, the fraction of ligand bound, so we may write

$$F_{rel} = F_f + \left(\frac{F_b - F_f}{F_f}\right)(F_f)(f)$$

$$F_{rel} - F_f = (F_b - F_f)f$$

*The fluorescence yield, the intensity relative to a standard, will be discussed further in Sec. 7.6.

So we have
$$f = \frac{F_{rel} - F_f}{F_b - F_f} \tag{7.9}$$

Given f, we can find $[PL]$ from the simple relation

$$[PL] = f[L_0]$$

and therefore we can calculate $\bar{n}$:

$$\bar{n} = N\phi = N \frac{[PL]}{[P_0]}$$

We know $[L_0]$ and $[P_0]$ from initial solutions and dilutions, and $[L]$ is given by $[L_0] - [PL]$.

So we see that the binding equilibrium is simply related to measurable fluorescence parameters. In cases of strong enhancement of ligand fluorescence on binding it is possible to ignore F_f, while in cases of strong quenching of ligand or protein fluorescence on binding, F_b is negligible. In cases of the binding of more than one mole of ligand, the above equations assume a constant fraction of quenching for each successive mole bound. This assumption must be checked experimentally, and if it does not hold true appropriate corrections must be made.

These calculations make several assumptions: (1) that the protein does not absorb at the wavelength of ligand excitation, so that the amount of light absorbed by the ligand will not vary as the ratio of protein and ligand concentrations varies; (2) that the protein does not absorb at the wavelengths used to observe ligand fluorescence, so fluorescence intensity is not dependent on the protein concentration in addition to the binding dependence, and (3) that the ligand does not absorb at the wavelengths used to observe protein fluorescence (when that is the intensity change being measured), so that fluorescence intensity is not dependent on ligand concentration in addition to the binding dependence.

The methods of measuring binding equilibrium by the change of a measurable parameter on binding have so far been limited to cases of fluorescence intensity changes. It is also possible to monitor binding by the change in *polarization* of a ligand whose intensity does not vary between the free and bound states, provided it is bound rigidly to the protein so that there is a large change in the rotational relaxation time (see Chapter 4). This method is discussed by Dandliker et al. (1964), Weber and Daniel (1966), and Deranleau and Neurath (1966). It is also possible to do polarization measurements in a case of combined intensity and polarization changes on binding (Rawitch and Weber, to be published).

It is also possible, in cases in which the protein fluorescence is quenched on binding, to monitor binding by the decrease in *lifetime* resulting from energy transfer from tryptophan to the ligand. For details, see Chapters 5 and 6.

7.6 Fluorescence Measurements

A. STANDARDS

Unlike absorbance measurements, fluorescence intensities are arbitrary numbers. The intensity is proportional to the number of emitting molecules, i.e., the concentration in any small volume of the solution, and the intensity of exciting light. But all small volumes of the solution do not receive the same intensity of exciting light, due to its absorption. Thus, fluorescence measurements are plagued with a problem referred to as *geometry*. It is a function of the path and the intensity decrease (absorption) of the exciting light through the cuvette, and of the small volume of the cuvette upon which the optics leading to the emission photomultiplier are focused. Small changes in geometry factors, particularly the absorbance of the solution, affect the intensity greatly. The intensity is also a function of slit widths, photomultiplier voltages, and amplifiction or attenuation steps preceding recording devices. Arbitrary amplification of the signal is a constant factor during one use of an instrument, and would not affect measurements during the same run which are to be compared to one another, as is the case for a typical binding study. But geometry changes due to changes in absorbance between different samples must be corrected for. The correction used is the comparison of each intensity to a standard.

A standard is a fluorescent solution of the same absorbance as the sample at the exciting wavelength, which therefore absorbs light and emits fluorescence in precisely the same geometry as the sample. Its fluorescence spectrum does not have to coincide with that of the sample, even though it must be measured at the same wavelength; it is permissible to measure intensity changes at any place in the band where there is appreciable intensity. The use of standards is particularly important in fluorescence binding studies, since a large range of optical densities is often covered in a titration. Any convenient standard may be used in protein binding studies. In the case of a ligand which undergoes fluorescence enhancement on binding it is often possible to use a solution of that ligand fully bound to some protein. The protein need not be the one under study if the ligand binds with acceptable affinity and fluorescence characteristics to some other protein, such as the much-used bovine serum albumin. The use of bound ligand as a standard has no advantage over any other good standard and is in fact more difficult to prepare and handle. Its main use is in those cases where no simpler fluorescent solution fits the requirements of a standard for the bound dye.

B. Evaluation of F_f and F_b

F_f and F_b were defined above as the fluorescence yields of free and bound ligand at a given concentration. By fluorescence yield we mean the intensity relative to some appropriate standard of the same absorbance at the exciting wavelength. This quantity is similar to the relative quantum yield since the sample and standard are absorbing the same number of photons in the same geometry, and their emission intensities are being compared. In this sense it is a number independent of concentration. However, the emission is not being integrated over the whole emission spectrum of each as in a true quantum yield comparison (see Chapter 5), but the two are being compared at a relatively narrow bandwidth at which the two emission spectra overlap, so the fluorescence yield depends on the emission wavelength and bandwidth.

To insure identical geometry, F_f and F_b must be measured at the same time as a titration or other experiment in which their values are needed. To determine F_f one simply measures the fluorescence of a solution of free ligand of any concentration and its standard and takes the ratio. The emission wavelength must be the same as that used in determining F_b and in all experiments in which these two values will be used. To determine F_b, one measures the fluorescence of a solution of ligand, of any convenient concentration, in enough excess protein that it is fully bound, taking its ratio to the fluorescence of a standard, as for F_f. However it may not be possible to make a solution of fully bound ligand of high enough concentration to give a good fluorescence signal, in which case one can use an extrapolation procedure in which the fluorescence intensity (relative to the standard) is measured for a series of solutions of the same total ligand concentration but increasing protein concentration, resulting in an increasing fraction of ligand bound. The results can be extrapolated to infinite protein concentration at which all the ligand would be bound, by plotting $1/F_{rel}$ vs. $1/[P_0]$ (see Fig. 7.7). In actual practice one may save protein and avoid making several solutions by beginning with the most concentrated protein solution and diluting it with a solution of ligand of the same total concentration.

If binding is to be followed by *polarization changes* instead of intensities, quantities completely analogous to F_f and F_b can be determined.

C. CONSTANCY OF FLUORESCENCE YIELD WITH SATURATION

An assumption in the equations of Sec. 7.5 is that the change in quantum yield (q) of a protein or ligand is the same for each molecule of ligand bound. The assumption may be easily tested using information from regular stoichiometry and equilibrium determinations. If the assumption is correct, addition of ligand to the protein under conditions in which it binds stoichiometrically will result in a linear relation between fluorescence intensity (relative to a standard) and $\bar{n}$, i.e., the initial part of a titration under stoichiometric conditions will be a straight line. Such a linear relation should be possible for any fluorescent system – quenching or enhancement of ligand fluorescence on binding, or quenching or enhancement of a covalent label or of inherent protein fluorescence, but it need not always be the case, and must be tested for any system studied.

If a high enough protein concentration for stoichiometric uptake of the ligand is not possible, one may do the addition under conditions of detectable equilibrium and correct for the free ligand. In this case it is simplest to deal with a nonfluorescent free ligand. We wish to correct the fluorescence for the absorption of part of the exciting light by nonfluorescent free ligand, i.e., to correct it to the value it would have if all the absorbed light had been absorbed by *bound* ligand. Since the fluorescence intensity is taken relative to a standard of the same absorbance, all values are also corrected for different ligand concentrations in different samples, so the relative yields of a set of samples are directly obtained.

The correction consists of simply multiplying the measured F_{rel} by the absorbance A of the total ligand at the exciting wavelength, divided by that of the bound ligand:

$$\text{Corrected relative yield} = \text{Measured } F_{rel}(A_{total}/A_{bound}) \quad (7:10)$$

For a derivation of this equation see the Appendix.

Different classes of binding sites may show different fluorescence changes on binding, causing the relative yield to change with saturation. It is also possible that some ligands which normally bind with fluorescence enhancement may in addition bind in nonfluorescent sites. If the total number bound in all sites ($\bar{n}$) is measured by some method independent of fluorescence properties, such as equilibrium dialysis, a decrease in yield with increasing $\bar{n}$ reflects energy transfer to the nonfluorescent ligands. Since the absorbance of bound ligand at the exciting wavelength cannot be measured independently of the total ligand, it must be calculated from the concentration of PL, which is calculated from the titration data. Or, since the ratio of absorbance is proportional to the ratio of concentrations, one may avoid separate absorbance measurements at each titration step and use $[L_0]/[PL]$, although this latter quantity may be less accurate than absorbance measurements.

Another criterion for a constant change in q (i.e., intensity) with $\bar{n}$ is the existence of an isoemissive point (analogous to an isosbestic point) in the fluorescence spectrum of protein and ligand. An isoemissive point is a wavelength at which the spectra for different $\bar{n}$ all cross. For equations showing the significance of the isoemissive point, see Anderson and Weber (1965, 1966). To determine whether there is an isoemissive point, one measures the fluorescence spectra of a series of solutions of protein and fully bound ligand of varying $\bar{n}$. It is best if the ligand is stoichiometrically bound, since it is difficult to correct the spectrum for free ligand absorption at all wavelengths.* It is also necessary to have a constant absorbance at the exciting wavelength in all samples, since it is not easy to devise something analogous to an intensity standard for spectra. In practice, if the absorbance of the protein is much greater than that of the ligand at the exciting wavelength, one may use a constant protein concentration and simply add successive increments of ligand. Thus a normal stoichiometric titration is done with the whole spectrum recorded on each sample, instead of just the intensity at one wavelength.

D. Some Practical Considerations

1. *Choice of Excitation and Emission Wavelength*

The absorption spectrum of a ligand often undergoes significant changes on binding to a protein. In general there will be one wavelength, the isosbestic wavelength, at which the spectra of the free and bound ligand, and any mixture of the two, all at the same total concentration, cross. Aside from considerations for fluorescence measurements, the existence of this feature is important to the

---

*Note the difference of this condition from the measurement of the isosbestic point, in which free ligand is permissible in the solutions of increasing $\bar{n}$.

interpretation of binding studies because it shows the existence of only two spectral forms of the ligand, bound and free, and all intermediate spectra must be simple sums of contributions from bound and free components. The absence of intermediate equilibrium forms in binding which are different from the fully bound species, such as $P + L \rightleftharpoons PL' \rightleftharpoons PL$, was assumed in the equations of Sec. 7.3, and must be experimentally demonstrated before those equations or methods based on them are used.

The isobestic wavelength is found by comparing the spectrum of free dye in buffer with the spectra as increasing fractions at the same total concentration are bound to the protein. This is done by measuring a series of difference spectra against protein, all at the same total dye concentration and increasing protein concentration, until no further change in the spectrum occurs by increasing the protein concentration. The spectra are best recorded on the same chart, with care taken that the wavelengths remain aligned. A typical result is shown in Fig. 7.4.

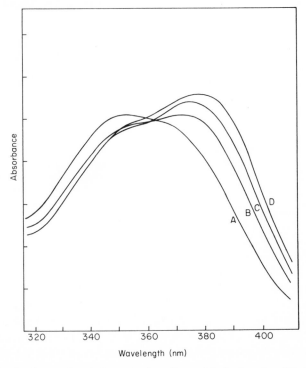

Fig. 7.4. Isosbestic wavelength determination. Curve A is the free dye, D is fully bound dye of the same concentration, and B and C are mixtures of bound and free dye.

In binding studies (and often in other fluorescence measurements) proper choice of excitation and emission wavelengths may allow some simplification of the system. The isosbestic wavelength is convenient for excitation because the absorbance at this wavelength is independent of the proportion of free and bound ligand, and dependent only on the total ligand concentration. Such a simplification makes it possible to make the standard by an addition or dilution titration exactly like that of the sample, beginning with solutions of the same absorbance.

In some cases the isosbestic wavelength may not be the most advantageous for excitation. In cases of inconveniently high absorbance of light at this wavelength for the dye concentrations needed to study binding, or in the case of a large red shift in the dye's absorption spectrum or binding, it may be best to excite on the long wavelength edge of the absorption band. In the latter case, the system can be simplified if there is a wavelength at which only the bound dye is excited. See Figure 7.5. In using edge-of-band excitation, the possibility must be kept in mind that errors may arise from impurities in the solution which could absorb a greater relative amount of light as the dye absorption is lowered.

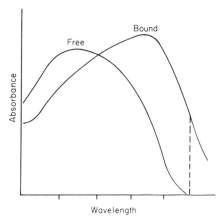

Fig. 7.5. Spectral change on binding in which the free dye alone may be excited.

The choice of wavelength for fluorescence emission measurements is usually not so critical as the absorption wavelength, though several considerations must be kept in mind. In cases of strong enhancement or quenching on binding (such as a several hundred-fold change in yield) it is usually possible to ignore the small fluorescence of the free ligand or quenched species, allowing the choice of any convenient wavelength and bandwidth. If the intensity is not measured for the entire emission band it is necessary to show that the change in intensity at the wavelength used is proportional to the change in area of the entire band. In using the F_{rel} obtained in the stoichiometric or equilibrium titration, one is generally measuring the intensity across a narrow part of the fluorescence band. If any shift in the peak wavelength occurs with $\bar{n}$, there will not be such a proportionality, but if a fairly large bandwidth is used, slight wavelength shifts of the peak wavelength will not be significant. The preferred method then is to use the large bandwidth.

It is desirable to avoid the short wavelength region of the fluorescence spectrum, which overlaps the absorption spectrum, since reabsorption of fluorescence occurs in this region, causing an anomalously low intensity. (See Chapter 5 for a complete discussion.) In cases where there is appreciable fluorescence both before and after binding, one must choose the wavelength of maximum difference between the two fluorescence spectra for the most accurate results.

2. *Some Problems with Inherent Protein Fluorescence*

As was mentioned, the quantum yield change may not always be proportional to the number of ligands bound. This is a particular danger for protein fluorescence. Several explanations may be given for nonidentical changes in protein quantum yield for each ligand bound. Unless each ligand is binding to completely identical subunits, the tyrosines and tryptophans in the environment of the sites may be arranged, and thus quenched, differently. And even if each ligand is binding on an identical subunity, if its quenching radius reaches into adjacent subunits the arrangement of the subunits may make the environment of two sites different. Another possible cause of nonlinear quenching is overlapping quenching volumes, so ligands bound after the first one already have a part of their quenching volume "occupied." This situation is illustrated in Fig. 7.6. Another obvious cause is the occurrence of some conformational change on binding, which might perturb previously identical environments to different extents.

A complete discussion of the complexities of protein fluorescence is beyond the scope of this chapter; an excellent review has been published by Weber and Teale (1965).

3. *Stability and Purity of Solutions*

A commonly encountered problem in binding studies is protein and dye instability. Vigorous stirring to mix solutions may promote destruction of the protein, especially at very high or low concentrations where aggregation and denaturation are more likely to occur. Protein

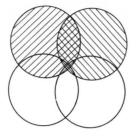

Fig. 7.6. Overlapping quenching volumes of two ligands bound on the same molecule, with two unoccupied sites.

adsorption to cuvette faces can be a serious problem at concentrations of about 10^{-6} to 10^{-8} M, where the amount adsorbed (and denatured) becomes a significant fraction of the total protein. Adsorption may be reduced by coating the cuvette with materials such as dichlorodimethyl-silane (Ferdinand, 1964). The presence of oxygen in the buffer may promote photooxidation under ultraviolet illumination, and it may be necessary to flush solutions with nitrogen and take care to record initial fluorescence intensities if changes appear to be taking place with time. If binding is being studied in a pH or temperature range of protein instability, special care must be taken that ligand binding is not promoting an undesired conformational transition, which could complicate results. (It is also necessary that all measurements be carried out at the same temperature, since temperature changes may affect both the equilibrium and the inherent fluorescence. Condensation on a cold cuvette may alter both the excitation and fluorescence intensity to a serious extent. A gentle stream of dry nitrogen directed on the cuvette face is an efficient method of removing condensation.)

Another common problem is fluorescent impurities in solutions or cuvettes. Cuvettes should never be allowed to dry with traces of solute in them; they should always be washed with an appropriate solvent just after use. If they are to be re-used immediately they may be rinsed in distilled water and air dried. Cuvettes should be stored in a cleaning agent, such as concentrated nitric acid, sulfuric acid plus dichromate or permanganate, or detergent (depending upon the manufacturer's recommendations). Cuvettes are best dried in a stream of clean air. Impurities in solutions may arise from contamination by untreated dialysis tubing, filter paper, corks, rubber stoppers and hose, stopcock grease, and various sorts of plastic containers and cuvette lids. Small residual fluorescence can be corrected for by subtraction of buffer blank intensities (at the same excitation and emission wavelengths) from each reading.

E. INDEPENDENCE OF BINDING FROM PROTEIN CONCENTRATION

If a solution of protein-ligand complex is diluted with a solution of liquid of concentration equal to the free ligand in the protein solution, the system will maintain the same value of ϕ if the degree of saturation is independent of protein concentration and dependent only on the free ligand concentration. The initial protein-ligand solution must be such that the free ligand concentration is readily measurable, i.e., not too close to 0 or $[L_0]$ (see Sec. 7.2).

Recalling Eq. (7.8),

$$F_{rel} = F_f[L] + F_b[PL] \tag{7.8}$$

we can introduce $\phi = [PL]/[P_0]$ and write

$$F_{rel} = F_f[L] + \phi F_b[P_0] \tag{7.11}$$

so if the binding is independent of $[P_0]$ we should obtain a straight line in a plot of F_{rel} vs. $[P_0]$ as we change $[P_0]$ by dilution at constant free ligand concentration [see Anderson and Weber (1965, p. 1953)].

If binding is found to be dependent on protein concentration, the system cannot be fully described by its binding equilibrium over a wide range of saturation at one protein concentration, but binding studies would have to be done at a number of protein concentrations in order to establish a binding surface.

F. MEASUREMENT OF LIGAND STOICHIOMETRY

A complete description of stoichiometry and equilibrium measurements, including the necessary conditions, parameters to be measured, and interpretation of data, was given in Secs. 7.2 and 7.4, but no attention was devoted to practical considerations of measuring the required parameters by fluorescence methods. The background necessary for understanding fluorescence measurements of binding systems has been discussed, and only a brief summary is necessary to relate fluorescence measurements more explicitly to actual binding parameters. Just as in the determination of F_f and F_b, the information on binding is essentially a simple series of fluorescence intensities or polarizations, from which one can calculate concentrations of the components of the system to determine stoichiometry or equilibrium.

The determination of stoichiometry requires a series of solutions at several different degrees of saturation of the protein, but with all the ligand stoichiometrically bound, to allow extrapolation to saturation with all the ligand in solution bound. Stoichiometry can be determined from plots of $[L_0]$ or $[L_0]/[P_0]$ vs. ϕ. $[L_0]$ and $[P_0]$ are easily determinable in a set of solutions made from the same stock solutions, and ϕ is easily calculable from Eq. (7.11), or the analogous equations for polarization or lifetime changes on binding.

G. MEASUREMENTS OF LIGAND EQUILIBRIA

To determine the equilibrium binding curve, we wish to make measurements covering the widest possible range of fractional saturation. The parameters required are the concentration of free ligand and ϕ. Equation (7.9) allows us to calculate ϕ again, and $[L]$ is calculated from $[L_0] - [PL] = \phi[P_0]$. The parameters $[L_0]$ and $[P_0]$ are known by known dilutions of initial stock solutions which are used to prepare the series of solutions of different saturation.

If one is exciting the samples at the isosbestic wavelength, the standard for each measurement may be obtained by diluting the initial standard in the same manner as the protein-ligand solution. To avoid serial dilution errors it is a good idea to start with a new initial solution and standard at several points along the saturation curve.

With fluorometers recording emission intensity directly, the sample and standard must be read in rapid succession to avoid lamp drift, but in instruments recording the ratio of incident light to emission intensity, lamp drift is corrected and the entire series of dilutions may be done on the sample and standard separately, so long as they are done essentially at the same time, to avoid changes in optics geometry.

H. MEASUREMENT OF LIGAND DISTRIBUTION

Measurement of ligand distribution among the protein population* can convey further information about the binding process. Simple binding results in a random distribution among all sites, while multiple binding results in a random distribution for each class of site, with the intrinsic K's determining the overlap between the distributions for each class of site. Cooperative interactions between sites on the same molecule can result in a bimodal distribution in which most protein molecules are either saturated or free, with intermediates, in much lower concentration. However, a bimodal distribution is not necessarily required for cooperative binding. Weber and Anderson (1965) point out that a normal ligand distribution results when cooperativity arises from

---

*"Ligand distribution among the protein population" refers to the fractions of protein molecules carrying $0, 1, 2, \ldots, N$ ligands. "Ligand distribution among the protein binding sites" refers to the order in which individual sites on the protein are filled.

a dependence of equilibrium among protein tautomers on $\bar{n}$. Measurements of ligand distribution among the protein population may be used to gain some information about binding processes in cases where the ligand affinity is too high to be measured by equilibrium methods, as in the demonstration of cooperative binding of heme to apohemoglobin (Teale, 1959; Antonini and Gibson, 1964). Distribution measurements have also been used to supplement binding equilibrium studies in demonstrating cooperativity in the binding of ANS to bovine serum albumin and particularly to demonstrate changes in cooperativity with saturation (Weber and Daniel, 1966; Weber, 1968). Ligand distribution may be inferred from two kinds of fluorescence experiments, observing either quenching of protein fluorescence by the ligand (apohemoglobin) or energy transfer among the ligands (bovine serum albumin).

Distribution may be inferred qualitatively from *protein* fluorescence quenching if three requirements are met: the protein must bind more than one mole of ligand, the quenching volumes of the ligands must overlap, and each ligand site, when solely occupied, must quench the protein fluorescence to the same extent. It is also helpful to make the simplifying assumption in calculations of a random ligand distribution among binding sites. If this is the case, an individual protein molecule that is a certain fraction saturated will be more than that fraction quenched, i.e., quenching is nonlinear with saturation within each individual protein molecule. (see Fig. 7.8). Now in the case of a completely bimodal distribution of ligand among the protein population, all the molecules are either fully quenched (fully liganded) or fully fluorescent (unliganded), and the degree of quenching will be linear with the amount of ligand bound. But in a case of random distribution (or an intermediate case in which there is a random component of the population) some of the molecules will be more fractionally quenched than fractionally saturated, and the quenching will proceed ahead of the saturation (see Fig. 7.8). Quenching can be followed by protein fluorescence intensity, but saturation must be measured independently. If the protein fluorescence does not overlap that of the dye, both may be measured by fluorescence.

Such experiments can in principle be quantitated, if one can calculate the limiting quenching for fully random binding, to determine if only a fraction of the population is binding with a random distribution (as in a search for intermediates in a cooperative binding process). This can be done by commonly used procedures involving the overlap integral between the donor and quencher (and usually again necessitates the simplifying assumption of a random distribution of ligand

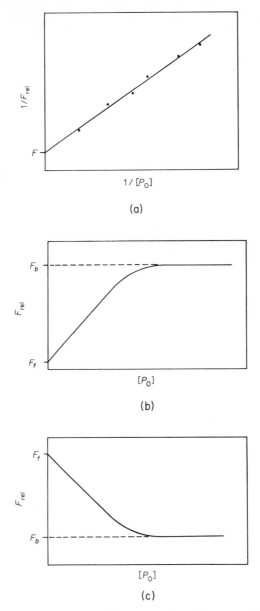

Fig. 7.7. Graph (a): Reciprocal plot of fluorescence intensity vs. protein concentration, at a fixed ligand concentration to determine the intensity at infinite protein concentration, where all the ligand is bound. Graph (b): A case of strong binding with fluorescence enhancement, in which stoichiometric uptake of all the dye occurs at obtainable protein concentrations. Graph (c): The same situation as (b), with ligand quenching on binding.

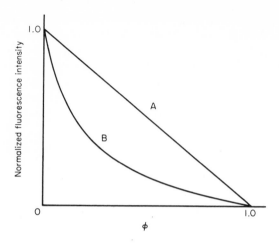

Fig. 7.8. Fluorescence quenching with ligand binding for a case of overlapping quenching volumes. A: Bimodal ligand distribution. B: Random ligand distribution.

among the protein binding sites). An example is a study of the binding of Fe^{3+} to transferrin (Lehrer, 1969).

As is the case with binding curves, one should measure quenching over the whole range of saturation. Deviations from normal behavior may occur near saturation, since any partially denatured component of the protein which might bind anomalously and with a lowered affinity would contribute most in this region.

Ligand distribution may also be measured by fluorescence polarization. Consider the case of a ligand which undergoes a large enhancement in fluorescence on binding, so measurements are weighted almost entirely in favor of the bound ligand population. If the ligands are rigidly bound, in the absence of transfer among them their polarization will be determined by the Brownian tumbling of the protein between absorption and emission. In a viscous solution (such as 50% sucrose) a high polarization can usually be obtained, if the ligand is excited in an absorption band whose oscillator is almost parallel to the emission oscillator. Then, in the presence of energy transfer, a significant decrease in polarization should be seen providing all the ligands in their different sites are not parallel to one another. Such polarization measurements are not possible for a case of ligand quenching on binding, but for this case or for a totally nonfluorescent ligand, if binding results in quenching of the protein fluorescence one could observe the polarization of the protein fluorescence in an analogous manner. If there are very few tryptophans and tyrosines in the protein, and they transfer energy among themselves, the system might be simple enough to give useful information. This sort of observation is complicated greatly by lack of understanding of the details of protein fluorescence. To simplify calculations, it may be necessary to assume random energy transfer among the tryptophans, which may not be the case, and the possibility for energy transfer from tyrosine to tryptophan may be difficult to account for correctly. The use of polarization of protein fluorescence to monitor any process in the protein is potentially a powerful method, but much independent knowledge is needed about energy transfer among the protein aromatic amino acids and its perturbation by any conformational changes on binding.

I. Measurement of Ligand Orientation

The depolarization resulting from energy transfer among bound ligands (see Chapters 4 and 5) might be expected to yield information on the relative orientations of the ligands, and on their orientation relative to the protein molecule. This has recently been shown to be true, and a detailed theory has been published (Weber and Anderson, 1969) with corroborative experimental work (Anderson and Weber, 1969). In the latter case ligand orientation has been inferred from changes in polarization with $\bar{n}$, the average number of ligand molecules bound.

REFERENCES

Adams, E. T. Jr., and Lewis, M. S. (1968). *Biochemistry*, 7, 1044.
Anderson, S. R., and Weber, G. (1965). *Biochemistry*, 4, 1948.
Anderson, S. R., and Weber, G. (1966). *Arch. Biochem. Biophys.*, 116, 207.
Anderson, S. R., and Weber, G. (1969). *Biochemistry*, 8, 371.
Antonini, E., and Gibson, Q. H. (1964). *Acta Biol. et Med. Germ.*, Suppl. III, 76.
Attalah, N. A., and Lata, G. F. (1968). *Biochem. Biophys. Acta*, 168, 321.
Bjerrum, J. (1957). *Metal Ammine Formation in Aqueous Solution*, P. Haase & Son, Copenhagen.
Boyer, P. D., and Theorell, H. (1956). *Acta Chem. Scand.*, 10, 477.
Brand, L., and Gohlke, J. R. (1966). *Fed. Proc.*, 25, 406.
Brand, L., Gohlke, J. R., and Rao, D. S. (1967). *Biochemistry*, 6, 3510.
Chen, R. F. (1967a). *Arch. Biochem. Biophys.*, 120, 609.
Chen, R. F., (1967b) in *Fluorescence: Theory, Instrumentation, and Practice* (G. C. Guilbault ed.), Marcel Dekker, Inc., New York, p. 443.
Chun, P. W., and Fried, M. (1967). *Biochemistry*, 6, 3094 (1967).
Churchich, J. E. (1967). *J. Biol. Chem.*, 242: 4414 (1967).
Dandliker, W. B., Schapiro, H. C. and Medeiski, J. W. (1964). *Immunochem*, 1, 165.
Daniel, E., and Weber, G. (1966). *Biochemistry*, 5, 1413.
Day, L. A., Sturtevant, J. M., and Singer, J. J. (1963). *Ann. N.Y. Acad. Sci.*, 103, 611.
Deranleau, D. A. (1964). *J. Chem. Phys.*, 40, 2134..
Deranleau, D. A. (1969). *J. Am. Chem. Soc.*, 91, 4044 and 4050.
Deranleau, D. A., and Neurath, H. (1966). *Biochemistry*, 5, 1413.

Edelman, G. M., and McClure, W. O. (1968). *Acct. Chem. Res.*, **1**, 65.

Edsall, J. T., and Wyman, J. (1958). *Biophysical Chemistry*, Vol. I., Academic, New York, Chapters 8–11.

Eisen, H. N., and Siskind, G. W. (1964). *Biochemistry*, **3**, 996.

Ferdinand, W. (1964). *Biochem. J.*, **92**, 578.

Fromm, H. J., *J. Biol. Chem.*, **238**, 2938.

Gally, J. A., and Edelman, G. M, (1965). *Biochem. Biophys. Acta*, **94**, 175.

Haugland, R. P., and Stryer, L. (1967). in *Conformation of Biopolymers* (G. N. Ramaschandran, ed.), Academic, London, Vol. 1., p. 321.

Hsu, R. Y., and Lardy, H. A. (1967). *J. Biol. Chem.*, **242**, 527.

Jaffe, H. H., and Orchin, M. (1962). *Theory and Applications of Ultraviolet Spectroscopy*, Wiley, New York, Chapter 20.

Klapper, M. H., and Klotz, I. (1968). *Biochemistry*, **7**, 223.

Klotz, I. (1953). In *The Proteins* (H. Neurath and K. Bailey, eds.), Vol. I, Part B. Academic, New York, p. 727.

Labeyrie, F., diFranco, A., Iwatsubo, M., and Baudras, A. (1967). *Biochemistry*, **6**, 1791.

Laurence, D. J. R. (1952). *Biochem. J.*, **51**, 168.

Lehrer, S., *J. Biol. Chem.*, (1969). **244**, 3613.

McClure, W. O., and Edelman, G. M. (1966). *Biochemistry*, **5**, 1908.

McClure, W. O., and Edelman, G. M. (1967). *Biochemistry*, **6**, 559 and 567.

McKay, R. H., and Kaplan, N. O. (1964). *Biochem. Biophys. Acta*, **79**, 273.

Millar, D., Minzghar, K., and Steiner, R. (1962). *Biochem. Biophys. Acta*, **65**, 153.

Nichol, L. W., Jackson, W. J. H., and Winzor, D. J. (1967). *Biochemistry*, **8**, 2449.

Parker, C. W. (1967). *Biochemistry*, **6**, 3408 and 3417.

Parker, C. W., and Osterland, C. K. (1966). *Clin. Res.*, **14**, 438.

Perkins, J. P., and Bertino, J. R. (1966). *Biochem.*, **5**, 1005,

Rawitch, A., and Weber, G. (1969). *Fed. Proc.* **28**, 413 (Abstract).

Steiner, R. F., and Edelhoch, H. (1962). *Chem. Rev.* **62**, 457.

Steiner, R. F., Roth, J., and Robbins, J. (1966). *J. Biol. Chem.*, **241**, 560.

Stryer, L. (1965). *J. Mol. Biol.*, **13**, 482.

Stryer, L. (1968). *Science*, **162**, 526.

Tanford, C. (1961). *Physical Chemistry of Macromolecules*, Wiley, New York, Chapter 8.

Teale, F. W. J. (1959). *Biochem. J.*, **25**, 289.

Theorell, H., and McKinley, McKee, J. S. (1961). *Acta Chem. Scand.*, **15**, 1811.

Velick, S. F. (1958). *J. Biol. Chem.*, **233**, 1455.

Velick, S. F. (1961). In *Light and Life* (W. D. McElroy and B. Glass, eds.), Johns Hopkins, Baltimore, p. 108.

Velick, S. F., Parker, L. W., and Eisen, H. N. (1960). *Proc. Natl. Acad. Sci., U.S.*, **46**, 1470.

Vestling, C. (1963). *Methods of Biochemical Analysis*, **10**, 137.

Weber, G. (1965). In *Molecular Biophys.* (A. Pullman and M. Weissbluth, eds.), Academic, New York, p. 369.

Weber, G. (1968). In *Molecular Associations in Biology*, Academic, New York, p. 499.

Weber, G., and Anderson, S. R. (1965). *Biochemistry*, **4**, 1942.

Weber, G., and Anderson, S. R. (1969). *Biochemistry*, **8**, 361.

Weber, G., and Daniel, E. (1966). *Biochemistry*, **5**, 1900.

Weber, G., and Laurence, D. J. R. (1959). *Biochem. J.*, **56**, 23P.

Weber, G., and Teale, F. W. J. (1959). *Faraday Soc. Discussions*, **27**, 134.

Weber, G., and Teale, F. W. J. (1965). In *The Proteins* (H. Neurath, ed.), and ed., Vol. III, Academic, New York, p. 445.

Weber, G., and Young, L. B. (1964). *J. Biol. Chem.*, **239**, 1415.
Winer, A. D., Schwert, G. W., and Millar, P. B. S. (1959). *J. Biol. Chem.*, **234**, 1149.
Winer, A. D., and Schwert, G. W. (1959). *J. Biol. Chem.*, **234**, 115.
Winkler, M. (1962). *J. Mol. Biol.*, **4**, 118.
Yoo, T. J., Ruholt, A. O., and Pressman, D. (1967). *Science*, **157**, 707.

Appendix: Derivation of the Equation for Relative Fluorescence Yield and Evaluation of Binding Data

A7.1 The Equation for Relative Fluorescence Yield

Equation (7.10) may be derived as follows: We wish to correct the fluorescence of a sample containing both free and bound ligand (fluorescence from bound ligand only) to the value it would have if all the light absorbed were absorbed by bound ligand. That is, we must correct by a factor f_{total}/f_{bound}, where f is the fraction of incident light absorbed. So we may write

$$\text{Relative yield} = F_{rel}\left(\frac{f_{total}}{f_{bound}}\right) \tag{A7.1}$$

We can show that, in general, for two absorbing species in a mixture:

$$\frac{f_1}{f_{total}} = \frac{A_1}{A_{total}} \quad \text{where } A \text{ is absorbance} \tag{A7.2}$$

Therefore, by simple substitution,

$$\text{Relative yield} = F_{rel}\left(\frac{A_{total}}{A_{bound}}\right) \tag{7.4}$$

We may derive (A7.2) as follows: Let n_1 be the number of photons being absorbed by species 1 in a length L, and n_L be the total number of photons being absorbed in length L. Then we may write

$$dn_1 = n_L \sigma_1 c_1 \, dL \tag{A7.3}$$

where σ is the effective cross section of the molecule, in which any photon striking is absorbed, and c is the concentration. We may also write

$$dn_L = - n_L(\sigma_1 c_1 + \sigma_2 c_2)\, dL$$

$$\int_{L=0}^{L} \frac{dn_L}{n_L} = - \int_{L=0}^{L} (\sigma_1 c_1 + \sigma_2 c_2)\, dL \qquad (A7.4)$$

$$n_L - n_0 = - (\sigma_1 c_1 + \sigma_2 c_2)L$$

where n_0 represents the number of photons *incident* upon the sample:

$$n_L = n_0 - (\sigma_1 c_1 + \sigma_2 c_2)L = n_0 - A_{\text{total}} \qquad (A7.5)$$

where $A = \sigma c_1$. Substituting n_1 into (A7.3) we get

$$dn_1 = (n_0 - A_{\text{total}})\sigma_1 c_1\, dL$$

and integrating:

$$n_1 = n_0 \sigma_1 c_1 L - A_{\text{total}} \sigma_1 c_1 L = A_1(n_0 - A_{\text{total}})$$

and likewise for a second species in the solution:

$$n_2 = A_2(n_0 - A_{\text{total}})$$

Hence

$$\frac{n_1}{n_1 + n_2} = \frac{n_1}{n_{\text{total}}} = \frac{A_1(n_0 - A_{\text{total}})}{A_1(n_0 - A_{\text{total}}) + A_2(n_0 - A_{\text{total}})} = \frac{A_1}{A_{\text{total}}}$$

This n_1/n_{total} may be equated to the f_1/f_{total} of Eq. (A7.2).

A7.2 Evaluation of Binding Data

A. Kinds of Binding Behavior

There are several types of binding behavior that must be understood before experimental results can be interpreted. More detailed descriptions are given in the references cited at the beginning of Section 7.2.

We have so far only considered binding to a set of equivalent and independent sites, a situation called simple binding. If there is more than one kind of site for a given ligand, the binding is called *multiple*. Multiple binding is described by a simple extension of Eq. (7.4a):

$$\bar{n} = \frac{N_1 K_{a1}[L]}{1 + K_{a1}[L]} + \frac{N_2 K_{a1}[L]}{1 + K_{a2}[L]} + \cdots \tag{A7.6}$$

with one term for each class of sites. Here, we have introduced $\bar{n}$, the average *number* of ligands bound, which is related to ϕ by

$$\bar{n} = N\phi \tag{A7.7}$$

where N is the stoichiometry of each class of site.

If any of the sites interact, so that binding at one site affects binding at another site (other than by simple steric blocking or electrostatic repulsion), the binding is called *cooperative* and is no longer described by the equations above. In positive cooperative binding ligand binding at one site facilitates binding at other, previously identical, sites, and in negative cooperative binding, ligand binding at one site makes binding at remaining sites less probable. Cooperative interactions are mediated through the protein itself, most probably by conformational changes.

Cooperative binding can be described by several equations, predicated on different models of the interactions in the protein. The most general equation is obtained by picturing the concerted equilibrium

$$P + jL \rightleftarrows PL_j \tag{A7.8}$$

where j is an *apparent* order of reaction, analogous to N in the simple equilibrium for N sites:

$$P + NL \rightleftarrows PL_n \tag{A7.9}$$

The parameter j is greater than 1 for cooperative binding, and less than 1 for multiple binding. From this equilibrium, we may write, analogous to Eq. (7.4),

$$\frac{\phi}{1 - \phi} = \bar{K}_a[L]^j \tag{A7.10}$$

where $\bar{K}_a$ is a *mean* association constant. The interaction parameter j can be determined from the binding measurements.

Other complex types of binding are possible; for example steric and electrostatic interactions, interaction between nonequivalent sites, and competitive binding, where some other species in solution competes with the ligand for a binding site.

B. STOICHIOMETRY

We have seen that stoichiometry may be determined by an *addition* titration, plotting the concentration of ligand added vs. the concentration of ligand bound (Fig. 7.2). In many cases, one of the axes may simply be a parameter proportional to the desired concentration. The break of the curve indicates N. It is necessary to extrapolate to the break point if the plot is curvilinear. To extrapolate accurately, as much of the plot as possible must be linear. It is easy for curvature of the plot to cause the initial portion to have a slope lower than the true stoichiometric line, even though it still appears linear. If both axes are normalized to $[P_0]$ the stoichiometric region will have a slope of 1, and can be easily seen.

A slight modification of the plot will allow the stoichiometric region to be seen even more clearly. One may plot $[PL]/[P_0]$ vs. the concentration of free ligand (Weber, 1965), or the equivalent $[L]/[P_0]$. From the derivation of Eq. (7.5) it is seen that $[L]/[P_0]$ is equal to $[L_0]/[P_0] - \phi$, and the new plot is essentially the same as the old one with the stoichiometric line now being the ϕ axis. All the points in the stoichiometric region will therefore fall on the $[PL]/[P_0]$ axis.

When the protein has several classes of binding sites of measurably different affinity, there will be a series of resolved "stair-steps" in the stoichiometric curve. But if the K_a are closer together than the limits of resolution, the curve may not level off in discrete steps, and the stoichiometry of each class of site may not be determinable.

Stoichiometry may also be measured by a *dilution* titration in which the initial conditions approach stoichiometric binding, so that N may be evaluated from the same plot as the equilibrium behavior. The details will be discussed in the next section.

C. EQUILIBRIA

It was pointed out in Section 7.2 that obtaining just N and the equilibrium constant is not the main point of a binding study. Rather, we wish to characterize the binding over as much of the binding surface as possible, and in fact we must determine the binding behavior in as much detail as possible for even an average N and K to be meaningful. We will discuss several kinds of plots which present binding curves, and the ability of each to show differences in binding behavior and give accurate values of N and K.

The simplest plot of binding data is similar to the one we have recommended for determining stoichiometry – plotting some parameter proportional to $\bar{n}$, the concentration of bound ligand, vs. some parameter proportional to [L], the *free* ligand concentration, as in Fig. A7.1. (Note that in the stoichiometric plot the abcissa was some

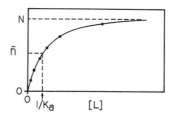

Fig. A7.1. Direct plot of binding behavior.

parameter proportional to the *total* ligand concentration.) This plot arises from Eq. (7.4a) and is analogous to the Michaelis-Menten plot widely used in enzyme kinetics (Dixon and Webb, 1964). The plot is a rectangular hyperbola for the simple case of equivalent and independent sites, and at high [L] the curve levels off to approach N on the $\bar{n}$ axis. K_a is given by the value of $1/[L]$ as $\bar{n} = N/2$, or by the slope as $[L] \rightarrow 0$, which is NK_a. Determining K_a from this plot is subject to considerable error, and it is extremely difficult to judge whether the binding is indeed obeying the assumption of one set of equivalent and independent sites.

A common method of evaluating binding data is by various plots which use a *linear* form of some binding equation. Several people, including Tanford (1961) and Weber and Anderson (1965), have made the point that much information is lost when the data are plotted to fit an equation that assumes a particular sort of binding behavior, because slight but significant deviations from that behavior may be at worst obscured or ignored as experimental error, and at best difficult to extract accurate binding parameters. With this approach the linear plots invariably attempt to extract both N and K from the same set of titration data.

Another very simple linear plot is based on the same equation for a single set of equivalent and independent sites (7.4a), which can be rewritten in the form

$$\frac{1}{\bar{n}} = \frac{1}{N} + \frac{1}{NK_a[L]}$$ (A7.11)

If one plots $1/\bar{n}$ vs. $1/[L]$, the equation is that of a straight line whose intercept on the $1/\bar{n}$ axis is $1/N$ and whose intercept on the $1/[L]$ axis is NK_a (Fig. A7.2). This plot is analogous to the common enzyme kinetics plot and has often been used in the literature. Deviations from simple binding behavior are not always readily apparent (Edsall and Wyman, 1958).

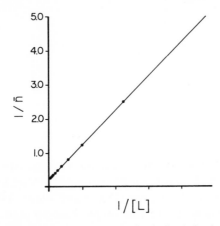

Fig. A7.2. Linear plot of simple binding behavior.

A somewhat better linear plot, again based on the equation for equivalent and independent sites, was first used by Scatchard (1949). Equation (7.4a) can again be rearranged to give

$$\frac{\bar{n}}{[L]} = K_a(N - \bar{n}) \qquad\qquad (A7.12)$$

which is the equation for a straight line if we plot $\bar{n}/[L]$ vs. $\bar{n}$ (Fig. A7.3). The intercept on the $\bar{n}$ axis is N, and the intercept on the $\bar{n}/[L]$ axis is NK_a. Multiple binding is readily seen as curvature to a less negative slope at higher $\bar{n}$, but evaluation of N and K_a for the cases of deviation can entail great uncertainty from extrapolations.

Many people have recommended the use of a *logarithmic plot* (Fig. 7.1) to present binding data, but it has been very rarely used in protein studies, even though it makes different types of binding behavior clearly distinguishable. Weber and Anderson (1965) point out that plotting a log [L] function instead of [L] excludes the region where the ligand is bound virtually stoichiometrically (where [L] is a very small number), and clearly shows the approach to saturation as a change in slope instead of relying on an extrapolation. The log axis

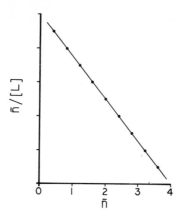

Fig. A7.3. Scatchard plot of binding behavior.

also makes it more convenient to represent the two or so orders of magnitude in $[L]$ that must be covered to go from 10% to 90% saturation of the protein. The form of plot proposed is $\log [L]$ or $-\log [L]$ vs. $\bar{n}$ or ϕ, analogous to the common representation of hydrogen ion equilibria. The suggested plot arises from Eq. (7.4a), which may be arranged to give

$$K_a[L] = \frac{\phi}{1 - \phi} \tag{A7.13}$$

written in log form as

$$\log\left(\frac{\phi}{1-\phi}\right) = \log K_a + \log [L] \tag{A7.14}$$

or in more general form as

$$\log\frac{\phi}{1-\phi} = \log K_a + j\log [L] \tag{A7.15}$$

This equation describes simple, multiple, or cooperative binding equally well using different values of j. The different orders of binding result in clearly distinguishable curves.

The form of the plot is sigmoid, as shown in Fig. A7.4, with the span on the $\log [L]$ axis depending on j. The span is defined as the length on the log axis between $\phi = 0.1$ and 0.9. For $j = 1$, simple binding, $\log L$ varies from $\log 0.9/0.1$ to $\log 0.1/0.9$, or $\log 9$ to $\log 1/9$,

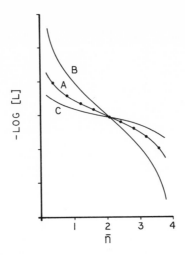

Fig. A7.4. Log plot of binding behavior. (a) Simple binding; (b) multiple binding, $j = 0.5$; (c) cooperative binding, $j = 2$.

which is 1.908 log units. For multiple binding the span is j times that of simple binding since, as we shall see, the curve is essentially a set of superimposed simple ($j = 1$) curves. For cooperative binding the span is j times less than that of simple binding.

The stoichiometry is given by the largest integral value of $\bar{n}$ approached by the curve. For multiple binding the curve shows stair-steps with the stoichiometry of each step given by its span on the $\bar{n}$ axis.

The K for each simple binding step is given by the free ligand concentration at the midpoint. For a cooperative curve, K is no longer the midpoint and may be found by fitting curves calculated for a given K to the data. Plotting $-\log[L] = \log 1/[L]$ gives $\log K_a$ directly, and plotting $\log[L]$ gives $\log K_d$ directly. It is difficult to resolve K's differing by less than a factor of about 10, although the increase in span becomes evident at smaller differences. A method for calculating K's for a multiple binding curve is given by Bjerrum (1957).

In principle, K can be determined from any mixture of P, K, and PL in which a measurable equilibrium exists, but Weber (1965) has shown that in fact the error in determining K is lowest at $\phi = 0.5$ and increases sharply on either side.

Author Index

Numbers in italics show the page on which the complete reference is listed.

Author Index

Subject Index

A

Absorption
anisotropy, 116
most probable transitions, 42, 43
processes, 37
relationship to emission, 26, 41, 150
Acetone, 32
3-Aminopyrene-1,6,8 trisulfonic acid, 79
Aniline, 31
Anilinonaphthalenesulfonate, 133
Anthracene, 34

B

Benzene, 20
Bimolecular complexes, 165; see Dimers
Binding, 92, 203
calculations, 216
equations, 206
equilibrium, 204, 208, 210
measurements, 92
spectral changes, 223
standards, 218
stoichiometry, 207
systems, 212
Bioluminescence, 56
Brownian movement, 92, 124

C

Cabazole fluorescence, 78, 82
Calibration of instruments, 153
Chemiluminescence, 56
Chromophores, 27, 31, 33

D

Deactivation pathways, 40
Depolarization of fluorescence 92, 97, 101, 143; see also Polarization of fluorescence
Dimers, 54
Dimethylaminonaphthalene sulfonate, 101, 103, 107, 110, 138
Dipole
angle between, 92
dipole interactions, 69
model of absorption, 88, 92

E

Einstein, 3
coefficient, 25
Electric field, 4, 5
Electronic, 4
dipole, 4
orbitals, 16, 17
transitions, 20, 27, 29
Energy transfer, 92, 127, 131
distance R_0, 136
efficiency, 139
in proteins, 131
practical applications, 141
quantitative theory, 135
summary of types, 134
values for, 138
Energy transitions, 11
Error of measurement
instrumental, 166
noninstrumental, 159